T0295851

Results-Based Systematic Operational Improvement

Results-Based Systematic Operational Improvement

Dr. Hakan Bütüner

CRC Press
Taylor & Francis Group
Boca Raton London New York

CRC Press is an imprint of the
Taylor & Francis Group, an **informa** business
AN AUERBACH BOOK

CRC Press
Taylor & Francis Group
6000 Broken Sound Parkway NW, Suite 300
Boca Raton, FL 33487-2742

Printed on acid-free paper
Version Date: 20161111

International Standard Book Number-13: 978-1-4987-2630-6 (Hardback)

Library of Congress Cataloging-in-Publication Data

Names: Bütüner, Hakan, author.
Title: Results-based systematic operational improvement / Hakan Bütüner.
Description: Boca Raton : Taylor & Francis, CRC Press, 2017. | Includes bibliographical references.
Identifiers: LCCN 2016043698 | ISBN 9781498726306 (hardback : alk. paper)
Subjects: LCSH: Operations research. | Continuous improvement process.
Classification: LCC T57.6 .B885 2017 | DDC 658.4/034--dc23
LC record available at https://lccn.loc.gov/2016043698

**Visit the Taylor & Francis Web site at
http://www.taylorandfrancis.com**

**and the CRC Press Web site at
http://www.crcpress.com**

Printed and bound in the United States of America by
Edwards Brothers Malloy on sustainably sourced paper

To my dear Mother, the only one who always
stands beside me under any circumstance.

Contents

Preface

Basically, I perceived the need for a systematic method of improvement. I wanted the method to be easy to understand and straightforward; based on fundamentals; and to be universally applicable to any type of business. Accordingly, I wanted to write this book in order to assemble the different ideas, processes, and techniques under the same roof, to develop a systematic methodology that is easily understandable and applicable.

Moreover, my intention was to bring a new perspective to the reader, and more significantly, to provide a different benefit by the application of this systematic methodology.

This book provides a complete set of practical improvement techniques. Where appropriate, I have refined these tools (from their original sources) to make them more user-friendly and effective. But, perhaps most importantly, you are guided in identifying under what circumstances you might use particular tools, and also how, and in targeting them directly to achieve effective results.

Systematic Improvement Planning consists of a pattern of six steps through which each project passes, and the three fundamentals involved in any improvement project. Each step is described in detail in the chapters.

This book aims to provide a detailed framework regarding systematic improvement planning. The method and its techniques, example cases, and working forms will guide you through preparing the improvement plan of your existing or future business.

- In Part I, the systematic pattern and characteristics of results-based improvement planning are overviewed.
- Part II describes the steps of systematic improvement planning in detail.

This book is written mainly for three groups of people:

- The first is the *individual* who desires to achieve improvement. He or she is skilled in devising improvement plans, but may not fully recognize that improvement planning of any business calls for a different procedure. His or her conventional approach must be replaced by broader analysis; his or her individual and factual analysis of specifics must give way to group opinions and evaluation of convenience or preference. This methodology presents an easy-to-follow procedure, and documents the work as it progresses. This will almost certainly ease the path to getting an approval.
- The second group is the *improvement project team*. By following a consistent approach, communications are greatly simplified, and the team will complete its project sooner with better results.
- The third group includes *small business owners*. They typically are reluctant (if indeed able) to commit large amounts of time and budget to improvement projects, but still have the need and desire to improve.

The contents of this book are drawn from the conclusions obtained from different application environments of the subject issue and by the composition of the cause-and-effect relations of these. To obtain maximum benefit from the book, the issues in the steps shall be addressed as a whole.

In reality, this book is an instruction manual. It has been designed to be specific, simple to understand, and easy to use. I hope it is of direct help to you.

Furthermore, I owe a great debt of gratitude to all my fellow businessmen who contributed through their opinions and efforts to the development of this book.

Most of all, I am very thankful to Richard Muther and his incredible innovation for planners, which is called Planning by Design (PxD), and is the best planning methodology that I have ever seen or heard of.

PxD is simply a methodology that lets you generate a methodology on any specific subject area. Based on PxD concepts, so many widely known methodologies have been developed for the use of planners in several businesses such as Systematic Layout Planning (SLP), Systematic Handling Analysis (SHA), Systematic Planning of Manufacturing Cells (SPMC), Multiple Careers Planning (MCP), Systematic Strategic Planning (SSP), and others.

Author

 **Dr. Hakan Bütüner, PhD**, is a graduate of TED Ankara College. He earned his BSc in industrial engineering from Middle East Technical University; MBA from Bilkent University, Ankara, Turkey; and PhD in engineering management from the University of Missouri–Rolla, Rolla, Missouri.

Dr. Bütüner has been active in both academic and professional fields for several years as a planning and programming manager and as a project manager overseas. He launched and served as the chairman of a new venture that aimed to bring various prominent international franchising concepts to Turkey.

Later, he worked as a strategic planning and business development director of Bayındır Holding, as a SPEED (operations and profit improvement program) country manager of Siemens Business Services, and as a general coordinator of Bell Holding. During the same period, he was also lecturing in various business schools and industrial engineering departments of Bilkent, Bhosporus, Bahcesehir, and Yeditepe Universities.

Dr. Bütüner is currently an affiliate of several U.S. companies involved in industrial management, engineering consulting, training, and software solutions. He is also the founder-president of the Institute of Industrial Engineers—Turkish professional chapter.

During his career, Dr. Bütüner has participated in several projects both in Turkey and other countries. He is a board member of the Institute of High Performance Planners in the United States. His other books are *Systematic Strategic Planning: A Comprehensive Framework of Implementation, Control, and Evaluation*; *Case Studies in Strategic Planning*; *Systematic Management: Why? How?*; *Competitive Advantages of Object-Oriented Business System*; *Database Technologies*; and *Customer Initiated Business System*. He has been honored by the decision sciences society Alpha Iota Delta.

PART I
IMPROVEMENT
PLANNING

The purpose of Part I is to provide information about productivity for businesses that have monitored its deficiency. It specifically provides definitions of results-based improvement and the systematic method with regard to improvement planning to raise awareness from the perspective of the employees.

1
RESULTS-BASED IMPROVEMENT

Various approaches are used for business improvement (Venge 2011). The following are among the most common ones:

1. *Quantum-leap improvement* is a common improvement approach adopted by businesses in the West. Improvement is perceived as an intermittent process performed in leaps and entailing major investments, as this process mostly bears the characteristics of renewal of technology and is performed by experts who mobilize all energy and financial resources of the business with respect to the solution of chronic problems.

    Major features
    - Short term
    - Quantum-leap improvement within certain intervals
    - Intermittent
    - Assigned to experts and therefore limited involvement
    - High investment required
    - Technology oriented

2. *Radical improvement* is an approach wherein improvement is an intermittent process performed mostly with the help of experts in a short time and with extensive participation, thereby mobilizing all energy and financial resources of the business to solve chronic problems. A continuous improvement culture has not yet been developed.

    Major features
    - Exciting within the defined period
    - Quantum-leap improvement within the short term
    - Temporary
    - Broad involvement

- Medium-size investment required
- Brainstorming oriented

3. *Results-based improvement* is an approach in which continuous improvement is viewed as a corporate culture within the business with the goal of complete elimination of problems through implementation of continuous improvement steps and the participation of all employees in the fields where the problems arise.

   Major features
   - Long-term and perpetual
   - Gradual
   - Continuous improvement
   - Organization-wide involvement
   - Minimum or medium-size investment required
   - Brainstorming and individual oriented

1.1 Productivity

Scopewise, productivity is defined in two different forms, though both are based on the same principles. According to its extensively recognized definition, it is "a rational way of doing the right thing in the right way and with an economic effort."

- Productivity is a progressive thinking that aims for continuous improvement.
- It supports the belief that "today is better than yesterday, and tomorrow should be better than today."
- It adopts economic and social lives to continuous changing conditions.
- It is an effort to apply new techniques and methods.
- It supports people development.

Since contemplating the productivity in such large dimensions precipitates identification and measuring issues, most implementers and researchers approaching the subject on the business level have favored using the concept of *business organization performance* instead of such a comprehensive productivity concept.

The second definition, which has never changed, is the classic definition (Coelli et al. 2005), "Reaching to the highest outcome with the lowest possible resource consumption." If any production unit has acquired more and better output than in previous periods with the mix of material, energy, machinery, workforce, and management resources used in that unit, this is interpreted as an increase in the productivity of that unit. According to this definition, "productivity is the outcome in the input/output relationships of all changes generated by the methods implemented in the existing manufacturing processes, input quantities, production capacities and output variations" (Coelli et al. 2005).

Productivity shows how successfully the factors of production are used in production or in the overall economy. The rates found by dividing the quantities or values of the outputs of production by the quantities or values of the inputs used for this production are considered indicators of the productivity level.

Today, competition extends beyond national borders and real competition in the international arena has gained impetus and significance. If developing countries cannot help their own businesses to improve their productivities and decrease their costs, they will lose all of their competitive possibilities against developed countries and remain in the vicious cycle of poverty.

The importance of productivity in increasing a nation's welfare is recognized by everyone. An increase in gross national income, based on an increase in labor effectiveness and quality rather than on the use of additional labor or capital, is very important. Accordingly, in case of distributing productivity gains according to their contributions, an increase in productivity directly enriches living standards. Changes in productivity influence rapid economic development, higher living standards, balance of payment, inflation control, and many other economic and social events to a great extent. These changes positively affect wage and salary levels, cost/price relationships, and capital investment and employment levels.

Competition and marketing strength of the economy are dependent on high and cheap production. High and cost-effective production, however, is a function of the productivity of factors used in production. Because of this, productivity influences the competitiveness of a country in the international market. Within the same goods-producing international market, if a country's labor productivity decreases then its competition becomes imbalanced. Additionally, if increased production costs are exactly reflected in the prices, customers will be more attracted to suppliers providing more cost-effective goods and the sales of national industries may decrease. Where high costs are not reflected in the prices and are covered by the industries, profits will also drop. This situation will lead to keeping production costs fixed by lowering production or wages.

In short, by productivity

- Employees work in better conditions and gain more in a shorter working period.
- Investors create new investment opportunities by generating additional resources.
- Manufacturers achieve higher gains at lower cost.
- Consumers find less expensive goods.
- Within a healthy economy, the country achieves rapid growth.
- In summary, society reaches a higher level of welfare.

1.1.1 Importance of Productivity for Businesses

Irrespective of the industry fields in which they operate, the primary purpose of businesses is to generate profit and share it while creating added value. A business needs to be productive to become profitable, and the major issue determining the relationship of productivity and profit is losses.

As measurement units referred to as inputs and outputs of productivity measurement are hard to calculate because they vary and are not intended for improvement purposes, *key performance indicators* are used to better focus improvement efforts on productivity goals (Parmenter 2010). Although key performance indicators provide some advantages in the calculation of productivity, they pose the following problems.

- They require financial data, which is often either unclear to the personnel in charge of productivity implementation or the personnel do not have sufficient information to interpret the data.
- Top management objectives are financial and it is hard to transfer them into technical levels as performance indicators.
- Most of the time achievements in performance indicators do not turn into financial gains.

- At times, even though key performance indicators leave the impression of a successful productivity effort, a loss may occur even with excellent productivity results. In fact, benefit/cost relationships are sometimes ignored.

As a result

- In using an academic and classical approach to productivity, one can derive several definitions and formulas. However, productivity itself is a result, and improvement efforts should be tailored in regard to root causes forming the results rather than the results themselves.
- Classical productivity measurements reveal the results created; however, they do not allow one to make future-oriented improvement plans.
- Engineers are action-oriented individuals and are interested in the root causes affecting productivity.

1.1.2 Relationship between Profitability and Cost

It is advisable to be aware of the relationship between *profitability* and *cost* in regard to variations occurring in the market, thus preventing undesirable outcomes. In other words,

- Profitability is a function of *cost* with a trend of declining *sales price* with regard to the competition conditions.
- Today, as a result of a global economy, sales price is not an issue determined by the management.
- Cost issues should be taken under control, and the information, systems, and organizational structures necessary for continuous reduction should be duly created.

Consequently, while examining the relationship between profitability and cost, bear in mind that a vast majority of the portion referred to as a quasi-cost consists of losses.

Therefore, the main aim is to be free from all wastes and losses burdening operations (George and Maxey 2004). The main operative strategy is, by increasing the speed, to reduce the duration of the stream and improve the quality, cost, and delivery performances. Accordingly, aligning material or information and value-added activities with customer requirements is very important. In short, the goal is to eliminate waste and loss in all operations, including

- Defects in products
- Overproduction
- Reactive quality controls
- Unnecessary materials handling
- Inventories of semifinished and finished products
- Unnecessary and non-value-added processes
- Employees on waiting time and unnecessary movements of workers

Therefore, remove the sources of waste and losses, such as

- Inefficient working methods
- Long setups
- Inefficient processes
- Lack of training
- Inefficient maintenance
- Long distances
- Lack of leadership

1.2 Results-Based Improvement

Considering the preceding discussions, results-based continuous improvement is a necessity for creating a productive and profitable environment that is free from losses. Improvement initiatives are intrinsically challenging efforts. Therefore, commencement of a successful improvement initiative requires the following conditions:

- Clarity of targets
- Clarity of time and budget
- A well-designed change process
- Most importantly, the active commitment and support of top management
- The top person in the organization serving as the active change sponsor
- Involvement and training of all employees to enable them to contribute
- Public recognition and reward of the desired new behavior

- A high level of respect for the change agents and drivers of change in the organization
- Assurance of communications throughout the organization from the top to the bottom, without allowing it to get lost in the middle

Improvements are necessary to

- Adapt to changes in strategies and goals, enter new business fields, be ready for business mergers, and so forth
- Perform efficiently in meeting the requirements of internal customers
- Gain competitive advantages
- Reflect technological developments
- Minimize variabilities

Results-based improvement requirements are

- Commitment and support
- Active participation of all senior management
- Patience and ability to persist
- Top-down commitment and involvement
- Common well-understood sets of metrics—defects and cycle time
- Goal setting
- Provision of required education
- Spreading the success story
- Sharing the rewards with those who contributed
- Using methods and necessary techniques for reaching success
- Developing the correct corporate culture
- Identifying the customers
- Setting stretch goals for reducing defects
- Cleaning up the obvious issues first
- Going for small as well as large improvements
- Encouraging continuous improvement
- Reporting progress at all levels

Basic tools for results-based improvement are

- Simple goals and strategies
- Simple management policies

- Common well-understood sets of metrics
- Necessary methods and techniques
- Uninterrupted organization
- Extensive training
- Teamwork
- Different wage incentive plans
- Visionary control
- Employee satisfaction surveys
- Communications management
- Prize-premium systems
- Recommendation systems
- White-collar employees in production
- Defined corporate values
- Reduction of layoffs

2

RESULTS-BASED SYSTEMATIC IMPROVEMENT PLANNING

Improvement in business is an operation in its own right. For this reason, a systematic approach ensures maximum benefit from improvement attempts. Results-based systematic improvement planning (SIP) is a structured approach to improvement. It draws on proven industrial engineering and quality improvement techniques, along with some of its own, and enables people to make positive changes. SIP is designed to help meet goals, solve problems, and implement ideas.

The aim of systematic strategic planning is to detect and minimize non-value-added processes and to provide a systematic procedure for continuous and satisfactory improvements. In other words, the purpose of this method is to provide teams with solutions to chronic and complex problems by

- Using a systematic approach
- Referencing problems into their root causes
- Selecting appropriate techniques for the problems encountered
- Developing alternative solutions
- Transferring the solutions into actions
- Tracking the financial and technical outcomes
- Standardizing solutions

In short, it can be defined as a broad and comprehensive method to help a business capture profitability and retain it permanently.

SIP can produce solutions to the following types of problems or expectations:

- Difficulty in meeting the demands of customers
- Ineffective and inefficient management of production planning and control processes

- Inability to measure the current production capacity properly
- Excessive time in the manufacturing process
- Low capacity utilization rates
- Low machine and labor productivities
- Inability to identify and/or eliminate non-value-added activities
- Inability to evaluate the current situation growth and investment fields and to set targets accordingly

SIP can be effective if

- There is a desire to consistently rank at the top on various productivity and cost metrics
- There is a desire to take a hard look at service operations, flow, equipment, and capacity. Although physical change is difficult and potentially costly, you would like to know to what extent the facility may limit your performance and opportunities to improve.
- Top management desires world-class performance, irrespective of rankings on current metrics. Defining "what is world class" may even redefine what services the firm should perform or cease to perform, and how it should perform them.
- Top management desires sustained and fundamental culture change—as a means to ongoing improvement, innovation, and world-class performance. This marks a significant departure from inherited existing culture of fuzzy adherence to established procedures.

Rather than focusing on strategic and organizational issues, it can be applied to operational problems, as illustrated in Figure 2.1, and in this way it can maximize the impact of efforts to

- Decrease non-value-added processes or materials
- Decrease business loss time
- Increase machine and facility utilization rates
- Decrease machine setups
- Improve customer due dates
- Decrease defect rates
- Decrease costs

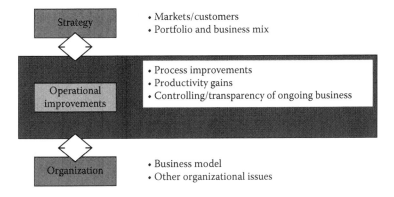

Strategy
• Markets/customers
• Portfolio and business mix

Operational improvements
• Process improvements
• Productivity gains
• Controlling/transparency of ongoing business

Organization
• Business model
• Other organizational issues

Figure 2.1 The SIP methodology focuses on operational improvements.

- Increase machine and labor productivities
- Decrease cycle times
- Initiate cultural change

By this method, the systematic and ongoing planning and development basis of a business will be established. As a result, a sustainable profitability and a stable competitiveness will be achieved. Table 2.1 shows real-life examples of the benefits of SIP. Monetary gains to be acquired may remain below intangible gains as a value, for example, increased confidence among employees, acquisition of new skills and greater energy.

The following are the basic advantages of SIP over other methods:

- Among all levels of business, it uses the same financial language for setting targets and tracking and evaluating successes.
- Selection of improvement techniques and implementation fields is based on financial losses within the business typology.
- Key performance indicators (which are related to financial gains) are used as indicators of the success of the implemented techniques, without incurring losses.
- By integrating implementations into daily activities and business processes, SIP is perceived as a set of routine activities by employees.
- SIP focuses not only on productivity and profitability improvement but also on eliminating root causes (losses).

Table 2.1   Examples of Results-Based SIP Benefits

	BEFORE	NOW
JOHN DEERE—BACK AXLES		
Lot size	2 weeks	1 week
Number of flow points	30	14
Total distribution distance (mt.)	225.000	78.000
Annual total distance (km.)	2.800	970
Trans. from Inventory (no. of parts)	38	3
The system pays for itself in six months		
TAYLOR MANUFACTURING—ICE CREAM PUMPS		
Lot size	30	1
Work-in-process	$ 30.000	<$4.000
Total distance moved (mt.)	5.000	50
Total cycle time	4 weeks	2.5 hours
Scrap (monthly $)	$1.600	$100
Labor productivity	90%	130%
CUMMINS MACHINERY—FLYWHEEL ENCLOSURE		
Labor productivity	42%	96%
Number of workers	49	20
Throughput time	2 days	1 hour
Decrease in inventory:		
Raw materials	6 days	1 day
Work-in-process	1.4 days	1 day
Finished goods	4 days	1 day
On-time delivery	30%	100%
Other improvements:		
Decrease in product cost	40%	
Decrease in materials trans.	40%	
Decrease in space	30%	

Source: Hales, H. L., and Andersen, B., *Systematic planning of manufacturing cells.* Dearborn, MI: Society of Manufacturing Engineers, 2002.

However, the success of SIP especially depends on

- Commitment of the management
- Acceptance of program targets
- Involvement of all management levels and employees
- Systematic control of implementation
- Responsibility for implementation at the management level
- Integration of project implementation in day-to-day business

Although SIP is used mainly on the shop floor of manufacturing companies, it can also be used to an even greater extent in many organizations outside the manufacturing realm: professional firms, sheltered workshops, hospitals, retail stores, and schools, to name a few—in other words, wherever people have a desire to get meaningful improvement results.

2.1 SIP: Approach

1. *Data gathering and analysis*: SIP begins with a clarification of which measurements are key with regard to business performance evaluation; then, data analysis is implemented in a way that will identify key variables and optimize the results.
2. *Focusing on processes*: Whether during design of goods and services, measuring performance, increasing productivity and customer satisfaction, or even while managing the work, SIP is a tool of achievement.
3. *Being proactive*: This includes setting stringent objectives and frequently reviewing them, clearly identifying priorities, not causing problems to emerge, and questioning why we execute the processes rather than defending them blindly.
4. *Infinite collaboration*: Immensity is intergroup cooperation toward achieving a single objective, such as offering value to customers. Immensity entails that efficiencies of a process in all phases are intertwined and need to be reviewed. It requires thorough understanding of the workflow in both the process and production chain and the real demands of end users.
5. *Eliminating variabilities*: In-depth evaluation of the variabilities helps you to understand the actual performances and processes of jobs. The goal is to reduce variability, which in turn means an extremely valuable improvement for many products, services, and processes.
6. *Targeting challenging goals*: Any business that focuses on a profitability objective through quantum leap rather than gradual improvement poses a significant challenge in the path of the improvement.

2.2 SIP: Three Fundamentals

The *fundamentals* of improvement planning—and therefore, the basis on which any improvement planning must rest—are Targets, Analysis, and Improvement Ideas, which are illustrated in Figure 2.2.

Targets is fundamental A. After the status of a business is determined by data analysis, measurable and quantitative improvement targets need to be set. For example, if a business is less profitable than the competition or the business needs ambitious targets, then the improvement target can be determined based on a set financial target such as Earnings Before Interest & Tax (EBIT). The total improvement target should be broken down into organizational units.

Analysis is fundamental B. The most critical point is to detect problems in the current status by a comprehensive assessment and to figure out where you may have the most leverage to take you to the targets.

Improvement Ideas is fundamental C. *Improvement Ideas* helps the organization to define how to eliminate the Leveraged Problems, which is brought about by matching fundamental A (*Targets*) with fundamental B (*Analysis*).

This fundamental characterizes the process of eliminating the causes of problems that the organization should follow to achieve the targets, and guides the organization about how it should be done. Every individual idea that turns into a project should bring a substantial improvement.

2.3 SIP: Six Steps

SIP is, therefore, a methodology that aims to determine these three fundamentals by following the six-step pattern, as shown in Figure 2.3.

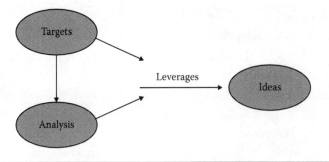

Figure 2.2 SIP—Three fundamentals.

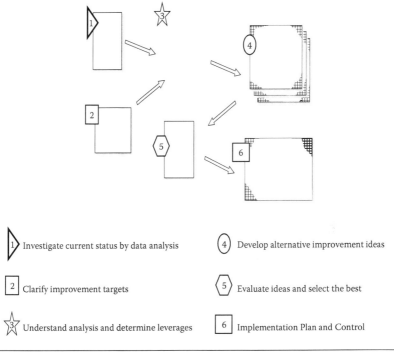

 Investigate current status by data analysis

2 Clarify improvement targets

3 Understand analysis and determine leverages

4 Develop alternative improvement ideas

5 Evaluate ideas and select the best

6 Implementation Plan and Control

Figure 2.3 Results-based SIP—Six steps.

Certain working forms (i.e., key documents and output) are used in applying the techniques in each step of the SIP pattern. The techniques to be used for improvement planning should not be perceived as a systematic procedure, and should be continuously repeatable and modifiable depending on the characteristics of individual cases.

2.3.1 Organize the Program and Investigate the Current Status by Data Analysis

Organize the program (which people are responsible for what and where they will work) and determine other related issues, such as external conditions, basic assumptions, or other related issues that need to be considered (such as legal, financial, technological, or organizational constraints), planning issues—who is responsible for what until when?, mandatory issues of the duty or the constraints under which the planning will be made, and content—planning dimensions or limits.

Additionally, get the facts, investigate the situation, survey the people involved, and document the findings. The data that are determined in this step are crucial for setting a reference base for the second step. This comprehensive study's main output is, no doubt, to provide an overall clear picture of the current status of the business.

With the help of the following sample questions, extraction of the value-generating processes' maps can be achieved:

- Which processes are crucial and adding value?
- What are the products and/or services?
- How do processes flow within the business?

2.3.2 Clarify Improvement Targets

Creating financial and operational improvement objectives is based on customer evaluations and on methods by which efficiency and competency can be sensitively measured. Determining clear, measurable, and quantitative improvement targets is necessary for developing meaningful measurements.

2.3.3 Understand Analysis and Determine Operational Leverages

While the second step determines the improvement targets, the third step concentrates on understanding how successful we can be to meet these targets under current conditions and how they can be sustained in the future by

- Determining the measurement and analysis basis
- Prioritizing the improvement areas and focusing on sources
- Determining the best operational improvement leverages

The main task is to analyze the 20% of the total cost of the business, the losses that correspond to the 60% of the cost of converting (the root causes that reduce the productivity), in order to select and implement the appropriate solutions for the chosen losses and to select the successful implementations with respect to their financial returns.

2.3.4 Develop Alternative Improvement Ideas

Develop and identify options, brainstorm the possibilities, refine them into viable options, and predict results for each option. With the help of in-depth analysis and creative thinking, develop solutions for the high potential improvement areas.

The improvement ideas should be based on the best appropriate techniques that meet goals, provide improvement in working speed and cost performances, eliminate errors, and increase the productivities and capacities of processes. Techniques and tools can be used both in solving complex problems and in valuing relatively simple improvement opportunities.

2.3.5 Evaluate Ideas and Select the Best

Compare and decide, evaluate the comparative results of the options, select the best option(s), and secure acceptance and approval. In this step, potential improvements are evaluated according to their potential and cost effects. Here you select the improvement ideas most suitable for your business. To do this, you evaluate the alternative improvement ideas based on quantitative and qualitative factors.

Also, alternative improvement ideas should be evaluated according to the risk factors that may be affected with their probability of occurrence. Therefore, both by evaluating according to quantitative and qualitative factors and considering possible risk issues, the best improvement ideas are selected.

2.3.6 Implementation Plan of Selected Improvements and Control

Act and follow-up, schedule installation, install, and audit the actual results. In short, all the effort is directed to implementing new solutions effectively and gaining measurable and sustainable benefits. The purpose herein is to reveal permanent implementations that will enable regular measurement, evaluation, and renewal of the projects, providing expected improvements.

In Part II, the six steps are described in detail.

PART II
Steps of Systematic Improvement Planning

Chapters 3 to 8 take us through the SIP pattern. They describe step by step the procedure and improvement techniques to use. Each chapter is devoted to a specific step of the pattern.

3

Organize the Program and Investigate the Current Status by Data Analysis

In this step, the operational environment of the organization is evaluated in general terms and the organizing program (which individuals are responsible for what, and where they will work) and other related issues are determined. *External conditions* are determined—basic assumptions or other related issues that need to be considered (such as legal, financial, technological, or organizational constraints). *Planning issues* is an explanatory stage—who is responsible for what until when, mandatory issues of the duty or under which constraints the planning will be made. *Content* is specified—planning dimensions or limits.

Additionally, gathering data regarding the current status and briefly analyzing them is the crucial action of this step.

Why you do it

This comprehensive study's main output is, no doubt, to provide an overall clear picture of the current status of the business. Determining the current status is necessary for setting measurable and quantifiable improvement targets. Additionally, before starting improvement work, an organizing program should be set up.

Deliverables
- Organization: Planning team list, duties, contact data, and limits
- Planning, resources, and budget issues that need to be solved before the program starts
- External conditions
- Program schedule
- Understanding the current situation

- Operational strategies, values, and policies
- Economies of scale
- Research and development, technology, and innovation capabilities
- Production planning and control system
- Effectiveness of operations control procedures
- Effectiveness of cost control system
- Basic financials
- Overall plant layout
- Materials handling plans and losses
- Utilities and building specifications
- Capacity utilization of labor and equipment
- Machine maintenance and setup efficiencies
- Review of appearance and cleanliness issues
- Productivity of labor and equipment
- Delays
- Inventory turnover rates
- Suppliers' performance
- Cost and availability of raw materials
- Organization chart
- Management performance
- Staffing charts
- Understanding of working hours, break times, staffing levels, and variations in work load and peaks
- Employee satisfaction and training levels
- Employee turnover rates
- Incentives to motivation
- As-is process flow charts
- Sales and production volumes
- Cycle times
- Customer complaints
- Defect types and rates
- Effectiveness of channels of distribution
- Aftersales service performances
- Products and their image and qualities

3.1 Improvement Program Organization

The success of the methodology requires the involvement of the following improvement program team members (with described duties).

Team members should include people who are most likely to be directly affected by the improvement program results and are capable of making improvements. Decide who will be the program sponsor, program manager (facilitator), program office members, and unit managers. It's also a good idea to list contact information for everyone.

The program sponsor

- Is overall responsible for the success of the program.
- Provides necessary resources.
- Reports the progress of the program to the board.
- Is the implementation coordinator.
- Is the escalation and decision authority in all questions regarding program direction, scope, and priorities.

The improvement program manager

- Drives the progress of the program.
- Is responsible for splitting the targets with management.
- Is responsible for measurement, analysis, and determination of operational leverages.
- Is responsible for the generation of ideas and the definition of improvement plans together with management.
- Monitors progress of implementation and reports the progress of the program to the management.
- Has a coaching function for improvement action coordinators.
- Is responsible for best-practice sharing with action coordinators.

The program office

- Provides the SIP method, techniques, and know-how.
- Supports the execution of the program actively, for example, hotline support.
- Supports cross-action exchange of experiences.
- Provides systematics for efficient implementation control.
- Plans and performs organization-wide communication activities.

The unit manager

- Coordinates the generation of ideas and the definition of actions together with the program manager and program sponsor.
- Monitors the progress of the individual unit improvement program and achievement of unit targets.
- Initiates the generation of additional ideas whenever necessary.
- Escalates all questions regarding the program to the improvement program manager.
- Is responsible for best-practice sharing within the unit.

3.2 Planning Issues, External Conditions, and Program Schedule

Get the team together to discuss the program. Planning issues should be explanatory—mandatory issues, constraints, and proposed solutions. External conditions are determined—basic assumptions or other related issues that need to be considered (such as legal, financial, technological, or organizational constraints).

- What can make this improvement effort fail?
- What kind of obstacles could be faced during the execution of this program?

The team should develop a program schedule. Use the six-step SPI pattern as a guide. Estimate the time requirements for each activity listed in the schedule, as well as who will do what. Determine completion dates for each activity and list them. Table 3.1 shows the orientation worksheet (Muther 2011).

3.3 Review the Current Situation

Get the facts and investigate the situation. Spend two to three days working with different jobs and shifts in the operation. Survey the people involved. Analyze ideas, observations, suggestions, and data, and document the findings. Present and discuss with supervisors and managers. Finalize conclusions.

The data that are determined in this step are crucial for setting a reference base for the second step. This comprehensive study's main output is, no doubt, to provide an overall clear picture of the current status of the business.

Table 3.1   Orientation Worksheet

Prepared by: _____   Business: _____

Authorized by: _____   Date: _____

PROGRAM TEAM			L-LEADER M-MEMBER R-RESOURCE O-OTHER		HRS. REQD.	
NAME	ROLE	DEPARTMENT	PHONE		EST.	ACT.
			TOTAL			

PROGRAM ESSENTIALS

Objective(s)   _____
External Condition(s)   _____
Situation(s)   _____
Scope/Extent   _____

PLANNING ISSUES	Imp.	Resp.	Proposed Resolution	Ok by
1.				
2.				
3.				
4.				
5.				
6.				

Dominance/Importance Rating   Mark "X" if beyond control

PROGRAM SCHEDULE

NO.	ACTIVITY	WHO	HRS.REQD.	DUE DATE	STATUS
1	Determination of Status				
2	Determination of Improvement Targets				
3	Get Facts, Analyze, and Determine Leverages				
4	Challenge Alternative Solutions				
5	Option Evaluation				
6	Establish Implementation Plan and Control				

Reference Notes _____
RICHARD MUTHER and ASSOCIATES – 756

With the help of the following sample questions, development of value-generating process maps is required:

- Which processes are crucial and adding value?
- What are the products and/or services?
- How do processes flow within the business?

Therefore, to be more concrete, reviewing and understanding the current status requires the following actions to be taken briefly, which are organized in Table 3.2. Review and understand

- Operational strategies, values, and policies
- As-is process flow charts
- Organization charts
- Management performance
- Job descriptions and grades
- Employee satisfaction and provided trainings through surveys
- Employee turnover rates
- Incentives to motivation
- Staffing chart by location, day, and shift
- Working hours, break times, staffing levels, and variations in work load and peaks
- Facility plans and materials handling
- Facility, machine, and labor capacity utilization
- Machine and labor productivity
- Cycle times
- Delays
- Machine maintenance and setup efficiencies
- Appearance and cleanliness issues
- Inventory turnover rates
- Suppliers' performances and due dates
- Cost and availability of raw materials
- Customer complaints and due dates
- Defect types and rates
- Sales and production volumes
- Effectiveness of channels of distribution
- Aftersales service performance
- Products and their image and qualities
- Economies of scale
- Research and development, technology, and innovation capabilities
- Production planning and control systems
- Effectiveness of operations control procedures
- Effectiveness of cost control system
- Financials and ratios

Table 3.2   Business Status Worksheet

Prepared by: _____   Business: _____

Authorized by: _____   Date: _____

	Factors	Status	Remarks for Target Setting
MARKETING/ SALES	Customer complaints		
	Customer due dates		
	Defect types and rates		
	Sales volumes		
	Products and their images		
	Effectivity of channels of distribution		
	After-sales service performance		
FIN./ ACC.	Financials and ratios		
	Economies of scale		
	Effectivity of cost control system		
OPERATIONS/ TECHNICAL	Process flows		
	Delays		
	Suppliers' performances		
	Suppliers' due dates		
	Raw materials cost and availability		
	Inventory turnover rates		
	Products and their qualities		
	Defect types and rates		
	Production volumes		
	Cycle times		
	Working hours, break times, variations in work load and peaks		
	Facility plans and materials handling		
	Facility and machine capacity utilizations		
	Machine productivities		
	Machine maintenance efficiencies		
	Machine setup efficiencies		
	Appearance and cleanliness		
	Production planning and control system		
	Effectivity of operations control procedures		
	R&D, technology, and innovation		
MANAGEMENT/ PERSONNEL	Strategies, values and policies		
	Organization structure		
	Management performance		
	Job descriptions and grades		
	Employee satisfaction		
	Employee turnover rates		
	Incentives to motivation		
	Employee training		
	Staffing levels		
	Labor capacity utilization		
	Labor productivity		

+	Positive
	Neutral
–	Negative

4

CLARIFY IMPROVEMENT TARGETS

After determining the status of the business, financial and measurable-quantitative operational improvement targets need to be set. What do we want to achieve? How should targets be distributed within the organization? By which data can we describe the targets? These types of questions should be answered in this step. Targets should be

- Achievable
- Realistic
- Motivating
- Challenging
- Specific and integrative
- Defined in a timely manner
- Measurable

Why you do it

Determining clear, measurable, and quantitative improvement targets is necessary for developing meaningful measurements and accessing planned improvements.

Deliverables

Financial and accordingly measurable operational improvement targets for the business and its units.

Approaches and techniques used

The scope of financial and operational improvement targets is decided mainly by the top level management by considering the business status determined in step 1.

For example, a business financial target can be set by considering its low profitability against the competition and/or a business may need an ambitious Earnings Before Interest & Tax (EBIT) target by considering its status, as shown in Figure 4.1. To be more specific,

EBIT margins: Firm vs. competitors (in % of sales)

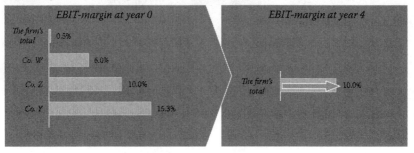

Figure 4.1 A business may need an ambitious financial target—example case.

Figure 4.2 can be used as a reference point for setting the business financial target.

Then, the business financial target should be converted into a total operational improvement target that then should be broken down into organizational units, as shown in Figure 4.3.

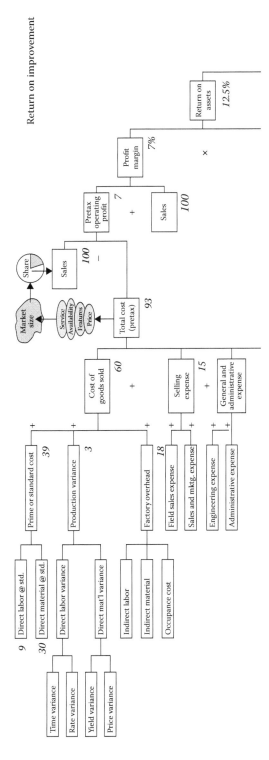

Figure 4.2 Useful form for setting a business financial target—example case. *Source:* Hales, H. L., and Andersen, B. *Systematic Planning of Manufacturing Cells.* Society of Manufacturing Engineers, Dearborn, MI, 2002. *(Continued)*

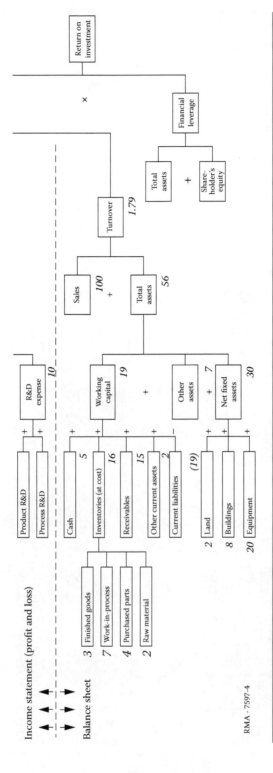

Figure 4.2 (Continued) Useful form for setting a business financial target—example case. *Source:* Hales, H. L., and Andersen, B. *Systematic Planning of Manufacturing Cells.* Society of Manufacturing Engineers, Dearborn, **MI**, 2002.

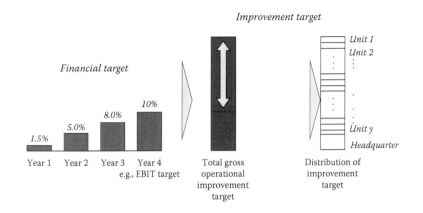

Figure 4.3 Distributing financial and improvement targets—example case.

5

UNDERSTAND ANALYSIS AND DETERMINE OPERATIONAL LEVERAGES

While the second step determines the improvement targets, the third step concentrates on understanding how successful we can be in meeting these targets under current conditions and how they can be sustained in the future by

- Determining the basis of measurement and analysis
- Prioritizing the improvement areas and focusing on sources
- Determining the best operational improvement leverages

Therefore, a comprehensive assessment is necessary to figure out where you may have the most leverage and where you may have gaps. For such an assessment, information on improvement needs and potential in the following dimensions of each process is necessary.

- *People and organization*: Roles, responsibilities, job definitions, training, incentives, performance measures
- *Policies and procedures*: Business rules and assumptions, controls
- *Physical facilities and technologies*: Capacities and capabilities

For this purpose, collect data on the end items, intermediate components, subassemblies, and parts for each final product to be made, including structure and configuration of major components and assembly sequence, preferably in an operation process or flow process chart in a value stream map. Understand customer ordering and consumption patterns and their variability, including the impact of a model mix. Identify the current triggers and information flows that drive production and movement between processes.

Therefore, to be more concrete, determining the best operational improvement leverages requires reviewing and analyzing problems (if any) in

- Cycle or processing times
- Changeover times
- Uptimes
- Scrap rates and yields
- Staffing
- Buffer size
- Process flows
- Information flows
- Management performance
- Employee satisfaction
- Employee turnover rates
- Working hours, break times, and variations in workload and peaks
- Facility plans and materials handling
- Facility, machine, and labor capacity utilization
- Machine and labor productivity
- Delays
- Suppliers' performance
- Customer ordering and consumption patterns
- Customer complaints
- Production volumes
- Products and their quality
- Research and development, technology, and innovation capabilities
- Production planning and control systems
- Operations control procedures
- Raw materials, work-in-process, and finished goods inventory policies and performances
- Relevance of operations

Afterwards, analyze findings and identify potential operational improvements. In this step, by using the results of measurements, various dimensions of critical losses affecting improvement targets of the business that we have set in the previous step are determined.

Therefore, problems (i.e., losses) should be defined clearly and in detail. The more clearly they are identified, the better and faster the solution process will be.

Why you do it

It is essential to gather all the necessary information about the situation you are trying to improve. Once you get the information, carefully document your findings in an organized way to figure out the key problems.

Deliverable

• Rated and identified causes for improvement issues

5.1 Measurement and Analysis Techniques

To solve the problems permanently, the causes rather than the symptoms of the problems should be eliminated. In this step, the measurements will be evaluated with respect to the employees, organization, procedures, facilities, technologies, and materials. An enormous variety of measurement and analysis techniques is available:

• Analytical quality tools
  • Pareto diagram
  • Fish bone diagram—cause and effect
  • Control tables
  • Histograms
  • Distribution diagrams
  • Control sheets
• Work sampling
• Work study
• Activity analysis chart
• Sequence diagram
• Customer surveys
• Financial analysis
• Process analysis
• Capacity analysis
• 5W
• …

5.1.1 Analytical Quality Tools

5.1.1.1 Pareto Diagram (for Root Cause Analysis)   A Pareto diagram is based on the *Pareto* principle, named after an Italian economist who recognized that 80% of the world's wealth was held by 20% of the countries. Applied to defect reduction efforts, this "*80–20*" rule becomes, 80% of the defects/problems are related to only 20% of all causes.

A Pareto diagram organizes data so that the vital few causes are separated from the trivial many. It is a very good visual tool for communicating information (see more in George and Maxey 2004). Figure 5.1 shows an example Pareto diagram.

Benefits
- Pareto analysis shows the results in graphical form and in this way major issues are revealed from the background.
- It provides a guide on the most important issues to address.

Type of defect	Tally	Total
Crack	ⅣⅣ ⅣⅣ	10
Scratch	ⅣⅣ ⅣⅣ ⅣⅣ ⅣⅣ .... ⅣⅣ //	42
Stain	ⅣⅣ /	6
Dent	ⅣⅣ ⅣⅣ ⅣⅣ ⅣⅣ .... ⅣⅣ ////	104
Gap	////	4
Hole	ⅣⅣ ⅣⅣ ⅣⅣ ⅣⅣ	20
Others	ⅣⅣ ⅣⅣ ////	14
Total		200

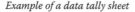

Example of a data tally sheet

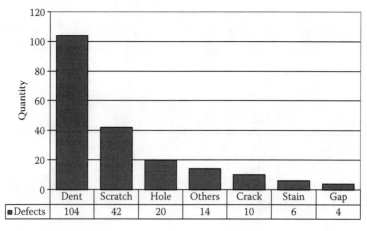

	Dent	Scratch	Hole	Others	Crack	Stain	Gap
∎Defects	104	42	20	14	10	6	4

Figure 5.1   Pareto diagram—example.

To construct a Pareto diagram

- Define which data need to be analyzed.
- Choose the categories by which the data will be sorted.
- Sort the data and determine
  - Total number of data points
  - Number of data points in each category
  - Percentage of total data points in each category
- Draw the horizontal and vertical axes of the diagram.
  - The left vertical axis represents the number of data points. Its scale must equal the total number of data points.
  - The right vertical axis represents the percentage of the total data points.
- The vertical bars representing the number of data points in a category are drawn starting with the category having the most data points against the left vertical axis. Then the next highest category, continuing until all categories, is shown.
- If it has been decided to use a category called *Other* to collect all the categories with very few data points, it is always shown against the right vertical axis.
- Plot the cumulative percentage from zero to 100% in line with each category bar.

To achieve continuous improvement, once the first cause is eliminated work on the next major cause, and then the next, and so on. In a Pareto diagram the purpose is to work on identifying the most leveraged problems in order to find the phenomenon constituting the *bad* situation. For instance, let us assume that typologies of the losses in TL currency are calculated within the business. The path to follow is to create breakdowns until the largest loss typology isolates the problem, which is shown as an example in Figure 5.2.

5.1.1.2 Cause-and-Effect Diagram   The objective is to show the route to the root cause of any problem in graphical form. A cause-and-effect diagram is most often used in improvement projects (see more in George and Maxey 2004).

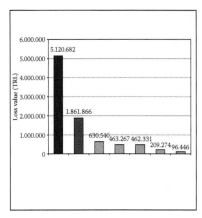

Losses Distribution Pareto

Which loss is the biggest?

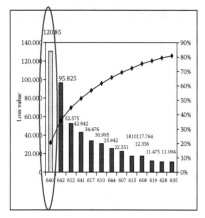

Fields Pareto

In which field, this loss is big?

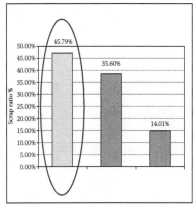

Processes Pareto

In this field, which process is big?

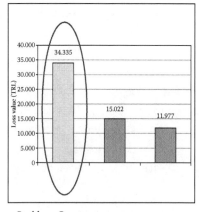

Problems Pareto

In this process, which problem is big?

Figure 5.2   Finding root problems through the use of a Pareto diagram—example.

Benefits
- Finding the main causes by dividing the problem into small pieces
- Strengthening teamwork
- Helping everyone to understand the factors that cause the problem
- Verifying the process template by being a roadmap
- Tracking brainstorming relationships

How to implement?

The effect(s), or results, are shown at the right side of the diagram. Causes fall into four main categories: materials and equipment (physical causes), methods (procedural causes), and people (human causes). Figures 5.3 and 5.4 shows typical cause-and-effect diagrams.

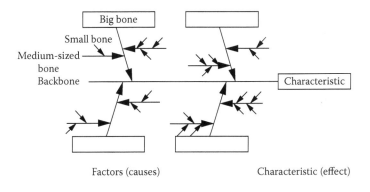

Figure 5.3   Cause-and-effect diagram.

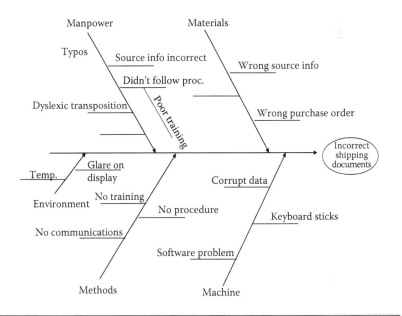

Figure 5.4   Cause-and-effect diagram—example.

Facts about causes are posted on the diagram along the branches leading to cause categories (often, lines are drawn from the cause to the branch; for this illustration, they were left off to reduce clutter). A variation of this diagram has the branches drawn at differing widths to show the relative total contribution of each cause category. This could, of course, be extended to include differing widths of lines to each cause:

- Big bones—categories
- Medium-sized bones—secondary reasons
- Small bones—root causes

5.1.1.3 Control Tables The primary purpose of a control table is to estimate the possible results of a product. Addition of a time element helps you understand the causes of variations in processes. A spreadsheet is a tabular graphic of data points centered in average data value and organized according to time range (see more in George and Maxey 2004). Figure 5.5 shows a typical control table.

Benefits
- Control tables enable prediction of the processes inside the designated limits and outside the control.
- They can be used to distinguish the causes of specific and detectable variations.
- They can be used for process control.

How to implement?
- Look at the upper and lower control limits. If your process is under control, 99.73% of all the data points will be within the lines.
- Upper and lower control limits represent the three standard deviations on either side of the mean.

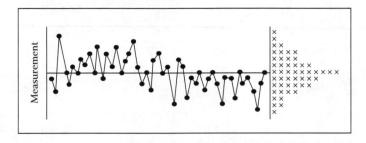

Figure 5.5   Control table. Unusual variation can hide in a frequency plot.

- Divide the distance between the centerline and upper control limit into three equal regions for representing standard deviations.
- Search the trend:
  - Two out of three consecutive points are in region C.
  - Four out of five consecutive points that are on the same side of the centerline are in region B or C.
  - Among ten consecutive points only one is in region A.

Figures 5.6 and 5.7 shows several example control tables.

Figure 5.6   Control table—example.

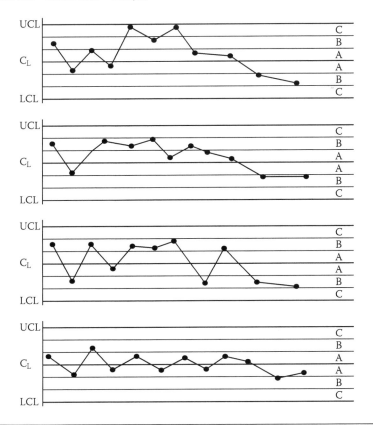

Figure 5.7   Control table—several examples.

The following strategies can eliminate the detectable variations of causes:

- Once you spot anything due to a detectable variation, investigate this cause.
- Replace the tools for compensating a detectable cause.

The following strategies reduce the common variations of causes:

- Do not try to explain the differences produced by a controlled fixed system.
- Fundamental changes on processes are required.

5.1.1.4 Histograms   The goal is to determine the distribution or variation of a set of data points in chart form (see more in George and Maxey 2004). Figure 5.8 shows a typical histogram.

Benefits
- Histograms make it possible to understand the variations of a particular process at a glance.
- The shape of the histogram will show the behavior of the process.
- The shape and size of the distribution reveals the sources of hidden variation.
- They help to determine the capacity of a particular process.
- They act as a starting point for the development process.

How to implement?
1. Collect data, 50–100 data points.
2. Select the data range.
3. Calculate the size of the class range.

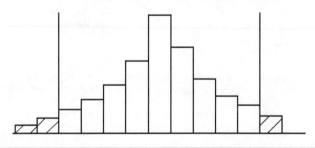

Figure 5.8   Histogram.

4. Divide the data points into classes and determine the class limit.
5. Count the data points in each class and draw the histogram.

5.1.1.5 Distribution Diagrams A distribution diagram shows a correlation between the two variables in a certain process. The data points have been distributed on a diagram. The goal is to uncover the correlation that may be present between the quality characteristics and the factors that may lead to them (see more in George and Maxey 2004). Figure 5.9 shows a typical distribution diagram.

How to implement?
1. Decide which paired factors to examine. Both can be measured on a progressive linear scale.
2. Collect between 30 and 100 paired data points.
3. For both variables, find the highest and lowest values.
4. Draw the vertical (y) and horizontal (x) axes of a graph.
5. Organize the data.

The shape that the point series will take will give you some clues regarding the relationship between two variables that you have tested.

5.1.1.6 Control Sheets Control sheets are used for collecting and editing the data so measured or calculated. The collected data can be used as information input for other quality tools (see more in George and Maxey 2004). Figure 5.10 shows an example control sheet.

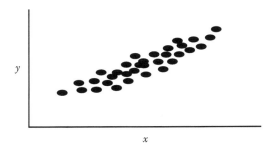

Figure 5.9 Distribution diagram.

		Machine 1	Machine 2
Operator A	Morning	X	X
	Afternoon	XX	XXXXXX
Operator B	Morning	X	XX
	Afternoon	XX	XXXXXXXXX

X = number of times the supervisor is called per day.

Defect type	Insufficient solder	Cold solder	Solder bridge	Blow holes	Excessive solder
Frequency	xxxxxxx	xx	xxx	xxxxxxxxxxxxxx	xx
Total	7	2	3	14	2

Figure 5.10   Control sheets—example.

Benefits
- Control sheets collect data in a systematic and orderly manner.
- They determine the sources of the problem.
- They ease the classification of data.

5.1.2 Work Sampling

Work sampling is a set of short-term observations that are based on random sampling for the purpose of recording frequencies of predetermined flow types in the same kind of work system (see more in Barnes 1980).

Obviously work sampling is just that—sampling work. At various times during a work shift the activity of each person is recorded as a code number, similar to taking a Polaroid snapshot of what the person is doing. Hundreds and even thousands of such instant observations are recorded over a period of many days or weeks. From these, a percentage of time spent on each activity is determined. It is important to understand that the activities are *not timed*. No stopwatches are used. An activity that takes a long time will simply show up more often (will be sampled more often), resulting in a higher percentage.

The activities of any one person are not important. What is important are the activities of the *whole group*, and how the mix of activities changes from month to month over the course of a year. The information is used to improve workflow and to schedule people properly.

Accuracy

How accurate is work sampling? After all, we are only observing activities at random instances. The statistical accuracy of a work sampling study depends mostly on how many observations are taken, both total and of a particular activity. With sufficient observations, work sampling is very accurate.

You, the subject

If you are a subject, a person being sampled, what should you do? Just what you always do. That includes such things as breaks and personal activities as well as your normal duties. It is almost impossible to influence a work sampling study by changing your individual behavior. Work sampling measures groups, not individuals, so just follow your usual routine.

You, the observer

If you are going to record activities of others as an observer, you will be given training. You will be given a list of coded activities and shown how to use an electronic data collector to record observations of those activities.

What will the results be used for?

The work sampling results will not be used to assess individual performance. In fact, we have set up the data collection so that every day the subject numbers are rotated and the daily number assignments are destroyed.

The results will be used to help a team of line station employees identify opportunities for improvement and provide management with statistical backup for their recommendations. For example, if the team found that an activity such as preparing paperwork for shipping parts took up 25% of an employee's time, they could look into a software solution to automatically generate the paperwork, perhaps reducing this to 15% of the person's time. Looking at these data together with other data being collected, the team could conclude that this recommendation could result in reducing service times to mechanics by an average of five minutes.

In addition, the results will be looked at hour by hour and compared with call volume. Since the line stations' activities are significantly driven by pushes, it will be very important for the team to look at recommendations based on busy times. In other words, when a push comes, what takes the most time, how can we make that faster, and what impact will that have on reducing service times and preventing delays?

How will we do this?

Having used this technique, the surveyor takes a tour within the business at predetermined times and observes flow types of specific work systems (for instance, *the machine runs* or *the machine stops*). A real picture of real flows based on several random observations can be attained by referencing to the work sampling technique's testimony for a myriad of goals.

A typical work sampling study is generally extended to a few weeks. Thus, if the results are to be really useful, meticulous planning is a prerequisite.

- The first goal of work sampling is to formulate and to select and describe the work systems to be observed. Work sampling is successfully used in identification of numerical indicators relating to the business such as mechanization level, utilization rate, loading rate, and workforce requirements; investigation of work flows in connection with the direction and planning of production; and identification of distributions of time.

- Then, to reach the desired results, the distribution of observed system elements to different flow types is determined. Flow types herein are generally divided in more detail than needed. In this way, a better overview can be obtained regarding the observed event.

- In the next step a tour plan is generated. This is a sketch-like representation of the observation points and sequences.

- The issue of the number of required observations is resolved by estimating the desired reliability level and share of flow type to be investigated most.

- At this point, the time for each round is set. What is important here is the random timing. Thus, both the statistical requirements are fulfilled and the results are not manipulated by the observer. Tour times are determined with the help of hour–minute–coincidence charts.

- After approximately 500 observations, generally an interim evaluation is conducted and whether the estimated number of required observation is appropriate or not is inspected.

For example, two supply attendants from each shift and location (one each day to provide coverage for seven days) will act as the observer for that shift. The data collector will be set up to beep randomly. The observer will do a tour and record what each worker is doing against a list of coded activities. No worker names will be recorded. A typical work sampling study form is shown in Figure 5.11.

5.1.3 Work Study

Time measurement consists of identifying work systems, specifically the working techniques, working methods, and working conditions and identifying individually the basic quantities, factors, performance degrees, and real times (see more in Barnes 1980). Their evaluation yields estimated times for specific flow periods. A typical work study is illustrated in Figure 5.12 as an example.

The key component of time measurement is the observation of a real situation by the person detecting the data. In the meantime, observation results are recorded. For this purpose, usually a time measuring tool and a form are kept in the hands of a data collector. The important thing is that time measurement should be repeated according to the information on the time measurement form. Conditions on which the measurement is conducted should also be recorded as meticulously as times.

Here, it should be set off from the following point: Once a time measurement is given to another surveyor, this person, with the information in hand, should be able to design a new work system yielding the results that can be compared with the emergent results in the observed work system. If this condition is met, it can be stated that time measurement reflects a picture of the observed work system, and thus it is repeatable.

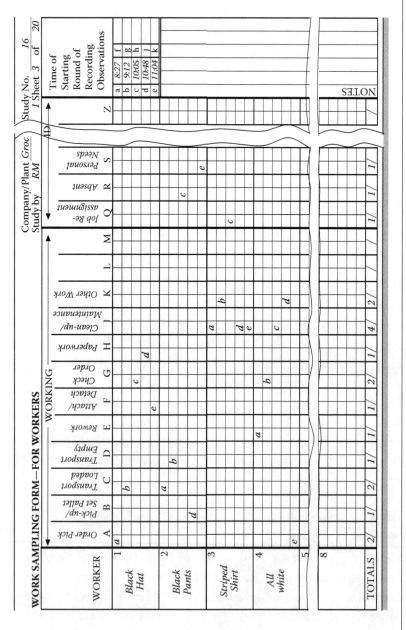

Figure 5.11 Work sampling form—example.

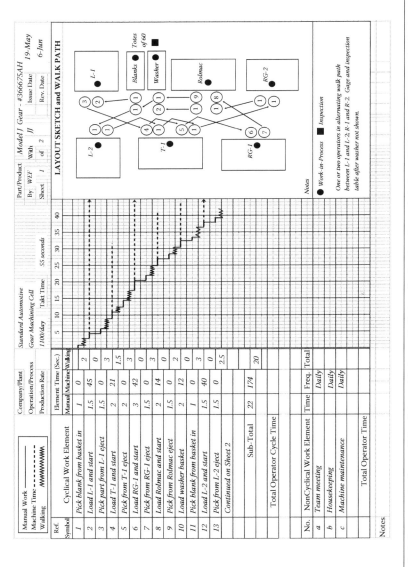

Figure 5.12 Work study—example. *Source*: Hales, H. L., and Andersen, B. *Systematic planning of manufacturing cells*. Society of Manufacturing Engineers, Dearborn, MI, 2002.

5.1.4 Flow Process Chart/Activity Analysis Chart

The Flow Process Chart is a classical industrial engineering tool. It was developed in the late 1920s by Frank and Lillian Gilbreth, and popularized in the 1930s and 1940s by Allan Mogensen in his Work Simplification Program of Continuous Improvement (Mogensen 1932).

Many improvements will center on a process. This may be a production process or an administrative process. Processes include a series of activities undertaken with a common purpose and may be classified as follows:

- *Operation*: An operation occurs when an object is intentionally changed in any of its physical or chemical characteristics, or is assembled or disassembled from another object. An operation also occurs when information is used or transacted.
- *Handling*: A handling occurs when an object is picked up or set down in conjunction with an operation or a transportation.
- *Transport*: A transport occurs when an object is moved from one place to another.
- *Delay*: A delay occurs to an object when conditions do not permit or require immediate performance of the next planned action. This includes queuing and buffers.
- *Storage*: A storage occurs when an object is kept and protected against unauthorized removal.
- *Inspection*: An inspection occurs when an object is examined for identification or is verified for quality or quantity in any of its characteristics.

Symbols for these activities are shown on the Activity Analysis Chart shown in Figure 5.13. The chart is completed by connecting symbols representing each activity in the process, along with a description of the activity. Preprinted symbols ensure that we identify all moves and record their distance and quantity or intensity of flow. Processing, delay, and storage times can also be recorded.

Built into the chart are classical analyses and actions that may improve the process. The purposes of the analysis section are

- What is the purpose of this process? Why?
- Where should this process be done? Why?
- When should this process be done? Why?

Flow process chart

Summary	Present		Proposed		Difference	
	No.	Time	No.	Time	No.	Time
○ Operations						
◇ Handlings						
⇨ Transportations						
☐ Inspections						
D Delays						
▽ Storages						
Distance traveled						

Plant _____ Sola Tech _____ Project _____ New Shaft Cell
Charted by _____ MCJ _____ Date _ 8/10 _ Sheet _ 1 _ of _ 4
☐ Man or ☒ Material _ Injection Pump Drive
Chart begins _ at Centering
Chart ends _ at Inspection

Details of method (☒ Present ☐ Proposed)	Symbols	Distance in feet	Quantity	Time	Analysis Why? (What? Where? When? Who? How?)	Notes	Action / Change (Eliminate Combine Sequence Place Person Improve)
1. Blanks in department stores	○◇⇨☐D▼		200				
2. To centering lathe queue	○◇◆☐D▽	45	25			Issue direct to lathe	
3. At lathe in queue	○◇⇨☐D▼		25	8 hr		Issue direct to lathe	
4. Into station	○◇◆☐D▽	15	25			Put in cell with subsequent operations	
5. Center and rough turn	●◇⇨☐D▽		1	0.92 min			
6. To contour lathe queue	○◇◆☐D▽	60	25			Put between centering and contouring lathes	
7. In queue	○◇⇨☐◆▽		25	4 hr			
8. Into station	○◇◆☐D▽	22	25			Put in cell next to centering lathe	
9. Turn to shape	●◇⇨☐D▽		1	0.95 min			
10. To engine lathe queue	○◇◆☐D▽	74	25			Put into cell with prior operations	
11. In queue	○◇⇨☐◆▽		25	4 hr			
12. Into station	○◇◆☐D▽	31	25				
13. Turn bearing ends	●◇⇨☐D▽		1	5.5 min			
14. To drop/pick-up area	○◇◆☐D▽	110	25			Eliminate storage with cell	
15. Wait for fork truck	○◇⇨☐◆▽		25	1 hr			
16. To milling department	○◇◆☐D▽	224	25				
17. In drop/pick-up area	○◇⇨☐◆▽		25	1 hr			
18. To department stores	○◇◆☐D▽	38	25				
19. In stores	○◇⇨☐D▼		163				
20. To department QC bench	○◇◆☐D▽	58	25	?		QC by cell operators	
21. In queue at bench	○◇⇨☐◆▽		25	4 hr			
22. Inspect (visual)/dimentions	○◇⇨■D▽		1	3.1 min			
23. In hold/release area	○◇⇨☐◆▽		67	3 hr			
24. To lathe department for rework	○◇◆⎄D▽	410	10%			Reduce defect rate Rework in cell	
25. To milling department stores	○◇◆☐D▽	58	90%			No stores; next operation in cell	

Figure 5.13 Activity analysis chart—example. *Source*: Hales, H. L., and Andersen, B., *Systematic planning of manufacturing cells*. Society of Manufacturing Engineers, Dearborn, MI, 2002.

- Who should do this process? Why?
- How should this process be done? Why?

The purposes of the action section are

- Eliminating unnecessary activity
- Combining or changing the place where a process is performed
- Combining or changing the timing or sequence of the process
- Combining or changing the person who performs the process

- Simplifying or improving the method, including the tools or machinery used

The quality of the problem is intended to be identified better, if it is evaluated with respect to the results of 5W1H analysis (see more in George and Maxey 2004).

5.1.5 Sequence Diagram

A sequence diagram shows the relationships of events (corresponding to activities) to time. It can also be used to show a correlation of events. A time scale is established, typically shown along the bottom of the diagram. Then events are posted. If an event must be preceded by a prior event, they are connected by a line.

In the example, events for control panels, invoices, and packing slips are shown in Figure 5.14. Almost immediately, the team questioned

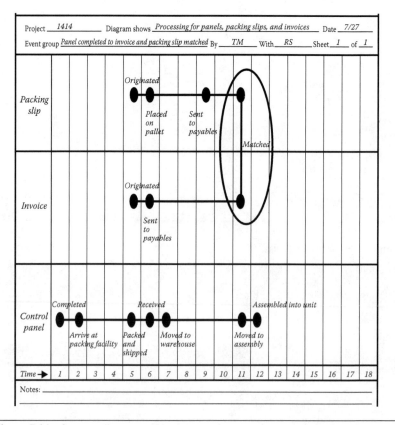

Figure 5.14 Sequence diagram worksheet—example.

why there were time lags associated with holding and storing the panels, and why there were similar time lags in sending packing slips to accounts payable.

5.1.6 Customer Surveys

These are studies conducted to obtain the information, experiences, and thoughts of various customers in relation to any problem or issue (see more in Hayes 2008).

The answers are evaluated according to a specific scale. The outputs such as cause-and-effect relationships and frequencies are obtained as a result of the analysis and they are used in evaluation.

Survey preparation process
- Determination of survey objectives
- Determination of survey variables
- Determination of sample pool
- Preparation of questionnaires
- Preparation of collection plans
- Collection of survey-related existing data in advance
- Determination of required sensitivity levels in estimation
- Determination of measurement techniques
- Preparation of the framework before the selection of samples
- Selection of samples
- Transferring data to electronic means and control
- Analysis of collected data and summarization

How will we do this?
- Identify your past, present, and future customers.
  - Use a specific customer only once.
  - Identify your current top ten customers.
  - Identify the top ten customers two years ago who are not major customers now.
  - Identify the top ten customers from five years ago who are not major customers now.
  - Identify the top ten potential customers (people who you'd like to have as your customers, but today are customers of your competitor).

- Identify ten customers who patronize your organization, spending a bit of their purchasing power for your products and services, but do not spend most of their purchasing power for your products and services.
    - This should give you a list of approximately 50 prospects.
- Identify your top five competitors.
- Hire a consultant. Those who do not reply should then be contacted by phone. A personal visit following the phone call may be necessary to get a minimum 50% response. This survey should be conducted by someone not associated with your organization and who should not at any time reveal to those interviewed that this survey is being conducted for your organization.
- Analyze the survey results. Identify those needs that are most important to your customers and then develop a marketing plan and internal strategy to meet those needs.
- Establish the frequency of the survey. This survey should be repeated regularly, preferably on an annual basis. By doing so, changes in the trends will become obvious.

The survey and tally sheets (ask identified customers and prospects the following questions):

1. Of the following characteristics, rate in order of importance to you. Rate from #1 through #4, with #1 being the most important, #4 the least important.

CUSTOMER NAME	#1	#2	#3	#4	ETC.
Quality of products offered					
Service offered to you					
Price of the product and service offered					
Delivery performance and speed of execution					
Total					

Add up the total scores for each of the four headings (quality of products offered, service offered to you, price of the product and service offered, and delivery performance and speed of execution) from the total survey. The one with the lowest score is the one that is the most important to your customer base. The one with the second lowest score is the second most important, and so on. Since you have all the original data, you can also review it by customer size, by distance from your location, and by any other criteria important to your business.

2. Who are your five biggest/most important suppliers?

1._____

2._____

3._____

4._____

5._____

Rate each one of the suppliers in the following four categories. Rate from #1 through #4, with #1 being the most important, #4 the least important.

NAME OF SUPPLIER	#1	#2	#3	#4	ETC.
Quality of products offered					
Service offered to you					
Price of the product and service offered					
Delivery performance and speed of execution					
Total					

When you analyze the results by totaling each of your competitor's total score, the results will give you a good idea of the competitive advantage each of your competitors has over you. Knowing this will greatly assist you in creating a strategy to offset those advantages.

3. Is there a product or service that we don't offer that you would buy if we did?

_____
_____
_____
_____
_____
_____
_____
_____
_____
_____

This survey will define products/services your customer would like you to provide.

5.1.7 Financial Analysis

Analysis of financial statements is the investigation of relationships among the items in financial statements and their trends in time in order to provide future-oriented estimations relating to the business; detect development aspects; and evaluate the financial status, results of activities, and financial development of the business (see more in Keown and Martin 2001).

What kind of information will be gathered through financial analysis?

- Distribution of business assets and the effectiveness of their use
- Distribution of business resources and utilization rates and benefits of their uses
- Short- and long-term loan payment capability
- Success level in the main activities of the business
- Business profitability
- Return on business assets
- Success in cash, inventory, and accounts receivable management

5.1.8 Process Analysis

Process analysis is the definition of the main process model and the identification of the improvement points related to the targets (as much as possible) (see more in Harrington and Esseling 1997). Activity-based process analysis is a kind of *optimization* purpose analysis intended for identification of processes adding direct value to work results, definition of activities, and enhancement of productivity that is based on process goals. For each process, deficiencies, problems, wastes, time losses, repetitions, and bottlenecks are investigated in detail.

Questions to be answered
- In the process, which activities add more value to the results?
  - Basic activities
  - Critical but not basic activities
- Are there any redundant steps that cause waste and time losses?
- Are there any repetitions and overlaps?
- Are there any deficiencies or inabilities in inputs?
- Can we reduce the control points and approvals?
- Do the process outputs coincide with the objectives?

As a result of the analysis, to keep business processes in line with the designated targets, the aim is to identify and refine unnecessary activities and in this way a base is formed for improvement suggestions. Typically the value-added period is 3% of the total time! (However, still most of the development efforts are focused on the value-added period.)

Documentation of processes
- Process description card (Figure 5.15)
- Main processes chart (Figure 5.16)
- Supplier–Input–Process–Output–Customer (SIPOC) (Figure 5.17)
- Process flow diagram (Figure 5.18)
- Functional flow chart (Figure 5.19)
- Value map (Figure 5.20)
- Spaghetti diagram (Figure 5.21)

Outputs
- Process relationship map
- Process simplification suggestions
- Necessary conditions for quick and productive processes
- Process responsibilities and authorities matrices
- Process roles and job descriptions

Process Name						
Sponsor of Process						
Mission of Process						
Process Class				Process Level		

Suppliers	Input	Output	Customers	Performance Indicators	Processes/Sub-processes/Activities

Sources:
Soundness of Process:
Soundness of Customer:

Figure 5.15   Process description card.

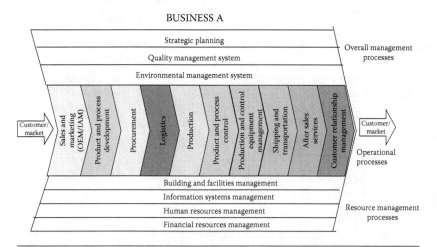

Figure 5.16   Main processes chart—example.

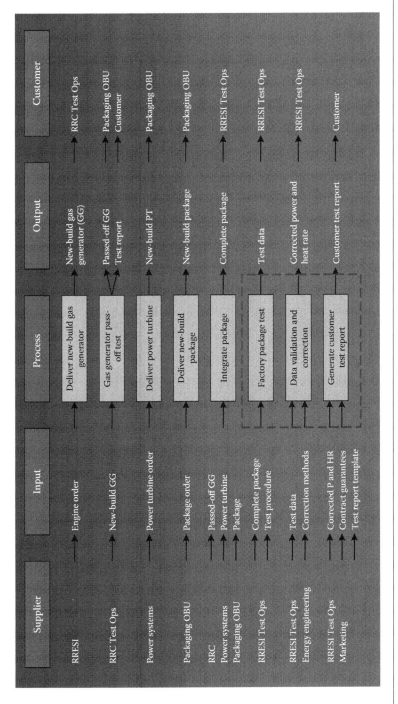

Figure 5.17 Supplier–Input–Process–Output–Customer (SIPOC)—example.

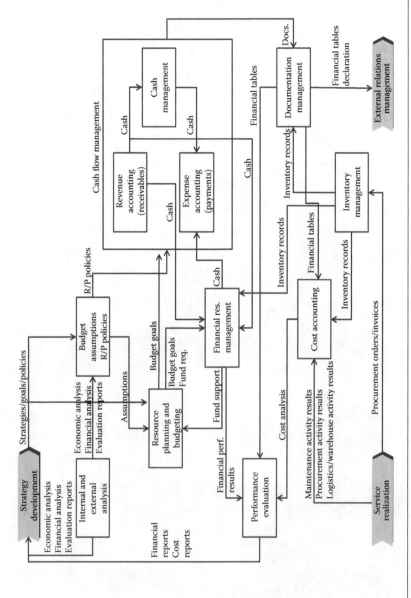

Figure 5.18   Process flow diagram—example.

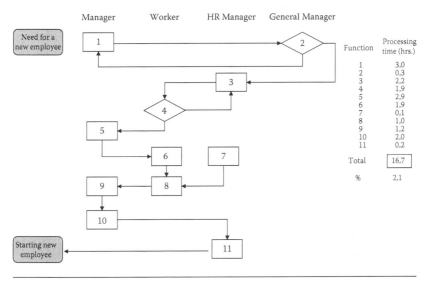

Figure 5.19 Functional flow chart—example.

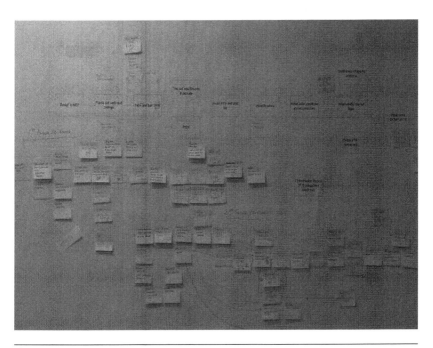

Figure 5.20 Value map—example.

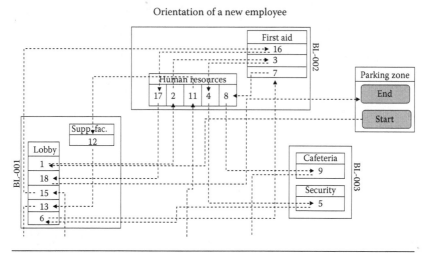

Figure 5.21   Spaghetti diagram—example.

5.1.9 Capacity Analysis

The purpose of capacity analysis is identification of the job capacities (see more in Hales and Andersen 2002). The priority is to produce the required accurate information for determining the process capacity. A time study should be conducted wherever it is necessary. As a result, the production capacity for each process will be designated.

To fulfill the processes, their necessary activities are identified and their times are measured and/or estimated. In this way, activity times per unit are obtained based on the roles in the processes. Working frequencies of the job are calculated and/or are estimated by linked them with the job results. Thus, necessary periodical workforces (for instance, monthly) are detected on the basis of the roles. Capacity requirements can be defined as the differences between the existing manpower resources and demand. Therefore, simplification, merging, and separation can be applied by comparison of capacity and work sizes.

Variables such as workforce, process error rates, activity durations, arrival and progress rates of the jobs to the system, costs and revenues, production goals, labor, and equipment utilization rates shall create a base for the improvement step. Capacity analysis investigates the following:

- The delay at each station
- The utilization rates of available resources

- The scenarios that can increase production with the available resources
- The impact of changes on all processes

As a result of the capacity analysis, the following outputs are targeted:

- Activities with simplification requirements
- Optimum activity finish time and capacity
- Capacity improvement alternatives

The planner uses charts and diagrams to visualize the routings for each class or subgroup of parts, and then calculates the number of machines and/or operators and workplaces that will be required to satisfy the target production rates and quantities.

When calculating the number of machines, planners must be sure to add allowances for downtime, schedule interference, and change-overs between individual and groups of parts.

Capacity analysis shows the types and quantities of machines required. A worksheet, with an example of this capacity utilization analysis, is shown in Figure 5.22.

A good capacity plan meets the desired production rate with an appropriate number of machines and level of utilization. Usually, the analysis will reveal over- and underutilization of some equipment planned. If the analysis reveals overutilization, the planner may choose to

- Remove parts from the cell to reduce utilization of the equipment.
- Purchase more equipment.
- Reduce process, changeover, or maintenance times.

If the analysis reveals underutilization, the planner may choose to

- Add parts to the cell to increase the utilization of equipment.
- Remove parts from the cell to eliminate the need for the equipment.
- Change the manufacturing process to eliminate the need for the implications of changes to product mix and peaks in production volume.

The Capacity Utilization Worksheet summarizes the machine processing hours for a group-of-parts cell making machined metal shafts.

CAPACITY UTILIZATION WORKSHEET   Project: Sola Tech. Shaft Machining Cell   Date: 9/2

Sheet 3 of 3
By: KS

| PRODUCT CLASS/DESCRIPTION | Work Center Number> | | Pieces per Year | PROCESSING TIME BY TYPE OF MACHINE | | | | | | | | | | |
	Part No.	Model		1 Center'g Lathe	20 Contour Lathe	30 Engine Lathe	50 Key Mill	12 Univ. Mill	40 Spline Mill	60 Thread Mill	70 Drill Press	80 Cylinder Grinder	90 Gear Mill	10 Gear Cutter
Carried forward from Page 2 (hours/year)				750	2,192	2,775	0	0	3,521	718	838	1,585	1,563	1,750
(b) Shaft with gears and threads														
Rear axle gear shaft	47345 53049	LS 20 B 30	300	8	55	28	–	15	–	15	–	123	15	–
Power take-off shaft	36456 70459	L 10 L 20	650	13	108	32	–	18	–	43	–	132	39	–
Auxiliary gear shaft	56097 78905 76890	LS 20 B 20 B 30	650	13	135	74	120	–	–	67	–	149	32	–

Figure 5.22 Capacity utilization worksheet—example.   (Continued)

	Part No.	Machine												
(a) Shaft with threads														
Injection pump drive shaft	46785	M 85	2,750	2,750	42	335	255	–	22	–	205	275	–	–	–
Steering knuckle arm	46056 45159 45907 45650 45432 45329	L 10 BH 10 L 20 BH 20 BV 20 B 30	2,750		46	160	225	–	168	–	182	463	–	–	–
King pin	46554 56354 10101	L–B 10 L–B 20 B 30	5,500	5,500	92	350	275	–	92	–	228	245	–	–	–
Machine Time (processing hours/year, incl. load and unload)					964	3,335	3,664	120	315	3,521	1,458	1,821	1,989	1,649	1,750
Setups/Changeovers (hours/year)					100	600	260	3	30	210	111	21	200	200	522
Planned Maintenance (hours/year)					24	25	28	6	14	33	20	180	50	52	360
Unplanned Downtime (hours/year)					12	12	15	3	7	16	10	90	24	25	180
Loaded Machine Time (hours/year incl. allowances)					1,100	3,972	3,967	132	366	3,780	1,599	2,112	2,263	1,926	2,812
Number of Machines Required					0.6	2.0	2.0	0.1	0.2	1.9	0.8	1.1	1.1	1.0	1.4
Number of Machines Available					1	2	2	1	1	2	1	2	2	1	1
Capacity Utilization					55%	99%	99%	7%	18%	95%	80%	53%	57%	96%	141%

Figure 5.22 (Continued) Capacity utilization worksheet—example. *Source:* Hales, H. L., and Andersen, B., *Systematic planning of manufacturing cells.* Society of Manufacturing Engineers, Dearborn, MI, 2002.

The bottom section of the worksheet summarizes the rough-cut analysis of machine requirements.

Allowances for setups, scheduled maintenance, and unplanned downtime have been calculated on worksheets such as that shown for the spline mill work center 40.

Given the use of averages and aggregate allowances, planners should examine machine utilization above 85% for possible overloading during extreme conditions. On this basis, five work centers in this example should be examined.

5.1.10 5W

5W is a technique in which possible causes of the problems necessary for the evaluation are examined (see more in George and Maxey 2004). For each possible reason, ask the "Why" question five times (as far as it can go), which is illustrated in Figure 5.23.

At this point, let us take a frequently given example, shown in Figure 5.24. Let us say that the problem is that a hanged frame frequently falls on the ground.

1. *Thesis No. 121*: The selected screw is not suitable for the frame weight.
2. *Validation*: The frame's weight is 3.5 kg. The screw can carry a 20-kg weight.
3. *Thesis No. 2111* will be handled as one of the root causes of the problem.

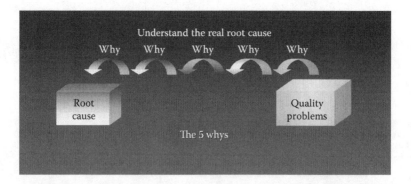

Figure 5.23 Five whys.

PROBLEM	1. WHY?	2. WHY?	3. WHY?	4. WHY?	PRECAUTIONS TEMPORARY	PRECAUTIONS PERMANENT
	1.	11.				
		12.	121.			
	2.	21.	211.	2111.		

Figure 5.24 Advanced five whys—example.

5.2 Determination of Operational Leverages

The need to improve implies that we consider the current situation as *bad*. What are the phenomena that make a current situation *bad*; in other words, what are the *problems?*

The primary causes of the problems need to be defined clearly with the support of the measurement and analysis techniques described in Section 5.1, that is, to figure out where you may have the most leverage and where you may have gaps.

The clearer the primary causes of the problems are identified, the better and the faster the solution process will be. In short, in this step, by considering the problems and their formation, various dimensions of leverage (primary causes for elimination of the problems) affecting business improvement targets that we have already identified in the previous step are determined.

For a reference, we can list some of the manufacturing-related problems as follows:

- Equipment related
  - Equipment breakdowns
  - Setups
  - Shortages in operators
  - Shortages in materials
  - Defective items
  - Reprocessing
  - Idle duration
  - Annual maintenance duration
  - Production planning
  - Maintenance in production duration
- Employee related
  - Due to equipment
  - Cleaning duration
  - Control duration
  - Maintenance duration
  - Testing and analysis
  - Managerial
  - Lack of automation
  - Complexity in processes

- Logistics
- Defective items
- Materials and energy related
  - Performances
  - Raw materials
  - Economic life
  - Maintenance materials
  - Leakage
  - Inventory
  - Consumption per unit
  - Defective items

In summary, bear in mind that profitability can be obtained at a lower cost, and toward this end, reduction of losses is needed. Our primary purpose is to eliminate the problems by being aware of the fact that the losses arise from the problems. Thus, to eliminate the problems (i.e., to find a suitable solution), recognizing the causes constituting the problems is of utmost importance.

5.2.1 Problems According to Their Appearance

Problems can be categorized into two groups: problems with and without determined root causes.

- *Irregular problems*: The problems that occasionally occur and deviate the loss rate from the initial or conventional level (even if not normal) and cause peak formation are considered irregular problems. If these problems are solved, losses return to the initial levels.
- *Chronic problems*: The problems that are considered *normal* because they exhibit continuity but in the meantime cause a higher level of losses than the ideal level. The ideal level is reached if these problems are eliminated.

Figure 5.25 illustrates the problems according to their appearance.

5.2.2 Determining Possible Operational Improvement Leverages

After measurements by the aforementioned techniques have been conducted and documentation of the measurement results with charts

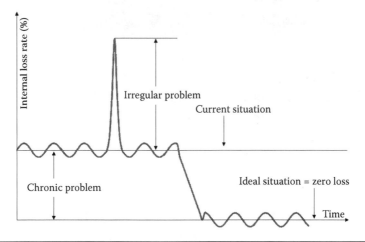

Figure 5.25   Problems according to their appearance.

such as activity and flow, sequence, cause-and-effect, and so on has been made, we are at the point of determining (i.e., analyzing and understanding) the problems according to their appearance by the use of three common techniques (line, bar, and pie graphs), which are shown as an example in Figure 5.26.

Both the line graph and bar graph, when used for a single item, readily show current levels and trends. When used for multiple items, they

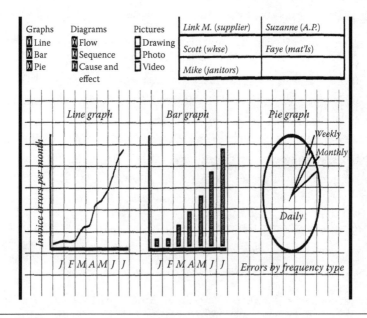

Figure 5.26   Determining problems according to their appearance—example.

may also show the distribution within each data point and trends for that distribution. The pie graph shows allocation, usually on a percentage basis.

Figure 5.26 also shows who, besides the program team, is involved in challenging the details of the measurements that are made. At the challenge meeting, visualizations are posted on a wall, along with copies of all measurement documents. The program team, along with others affected (the challengers), convene to review the documentation, discuss the findings thus far, and challenge every detail.

Consequently, problems that are determined according to their appearance (chronic or irregular) should be listed as a summary, as in Table 5.1.

Finally, during the meeting, a number of ideas should also be generated to figure out improvement leverages based on each defined problem, such as the one that is illustrated in Figure 5.27. During the identification of improvement leverages, a decision tree approach (see more in Skinner 2009) can be used. Consequently, identified leverages for each problem should be listed in Table 5.1.

Care must be taken to document each and every one of these for consideration in step 4. An appropriate wrap-up for the meeting is to review the leverages and ask every participant to add any others (if possible).

Table 5.1   Classified Problems with Their Leverages—Example Case

NO.	DESCRIPTION	TYPE (R – RANDOM, C – CHRONIC)	LEVERAGES
1	Invoice system	C	• Invoice processing procedure out of date • Copy of procedure not available to clerks • Procedure not considered "good practice" • Cost numbers printed on packing slips • Shipping report used to verify invoices • ...
2	Potential safety hazard with "popcorn" packing		• • • • •
3	Materials system assumes monthly delivery, which delays putting into warehouse		• • • • •

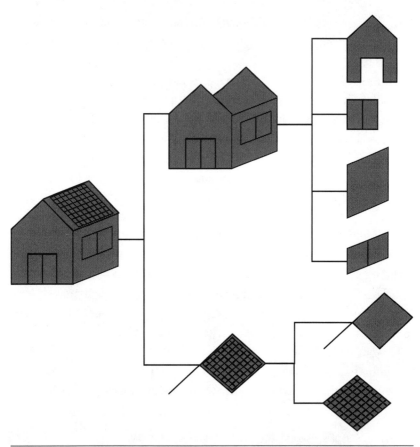

Figure 5.27 Decision tree approach for identifying improvement leverages. A decision tree breaks a problem down into smaller subproblems to identify improvement leverages.

6

DEVELOP ALTERNATIVE IMPROVEMENT IDEAS

The aim of this step is to differentiate the perspectives of the employees in production processes and to ensure achievement of the best results with the techniques used thereon by providing information related to the issues of improvement techniques whose deficiency has been observed within the business. Maximum participation of employees and of all white-collar staff is sought to the extent permitted by production.

The purpose of this step is to provide assistance for procedural, physical, and operational improvements and therefore for financial targets of the business to be met. When we have identified enough improvements by eliminating the leveraged problems, we should also be able to state whether the improved condition is *the best* that it could be. If we generate more than one solution to each leveraged problem, then the possibility of finding the most effective solution will most likely increase.

In practice, there are three common ways to arrive at better (lower cost and more efficient) solutions:

- Improve what you have; challenge each element of what you see and make a list of desired changes.
- Start with a clean sheet of paper and construct the ideal; look at the current state after you define the future state. Net the differences to arrive at a desired solution, once constraints have been addressed.
- Find the best practice wherever you can and copy or adopt it.

Why you do it

To move from a real situation to an envisioned (target) one, finding out what should be done by developing solution

77

options in regard to elimination of designated leverages (for fully eliminating the causes of the problem)

Deliverables

The potential aspects (with their value and volume drivers; and quantified present and desired situations) and costs of the recommended improvements for each determined problem as a whole or with its individual leverages

6.1 Process of Idea Development

The process of idea development is illustrated in Figure 6.1. An idea should point the way to improvement.

6.1.1 Generation of Ideas for Improvement

In this step, to eliminate the major causes of each problem (and therefore to solve the problem), the *candidate* solutions are determined by the various idea development approaches. Generate new ideas through

1. *Interviews*: The interview guideline helps to identify approaches for improvement. Rules for discussion:
   - Create a relaxed atmosphere (calm, no stress or time pressure).
   - Show you are attentive and interested.
   - Listen actively, and ask follow-up questions.

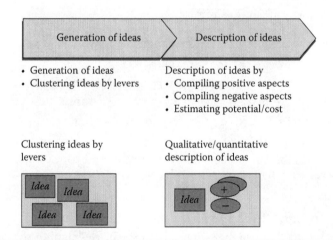

Figure 6.1   Process of idea development.

- The interviewer should never talk more than the interviewee.
- Adapt yourself to the situation and be flexible during the interview.
- Don't ask leading questions.
- Stay dispassionate and don't annoy the interviewee or make comments for social reasons.
- Give the interviewee the feeling that you are taking what he or she says seriously.
- Don't ask several questions as one; ask questions individually.
- Always ask if you don't know something.

2. *Idea workshops and brainstorming*: In brainstorming, the moderator introduces the topics and writes a point on the board (e.g., a question). He or she makes sure that everyone has understood the topic. The participants then call out their ideas one after the other, in detail where possible.

   In this creative phase there should be no criticism of the ideas presented, even if the comments are a joke, positive or negative. The moderator repeats the essence of the idea and briefly ensures that he or she has understood it correctly.

   One member of the group writes the idea on a Post-it® note and affixes it on the board. Ideas are collected in this way for as long as they keep coming and there should be no undue time constraint.

   The notes are then arranged and assessed, and the most important ideas are selected.

3. *Card survey*: The moderator also uses the card survey to present the topic, and writes a corresponding question on the board. Here, too, he or she makes sure that everyone has understood the topic. The participants are then asked to put their ideas on the topic in list form on the Post-it®s or on cards. The notes are then collected and arranged and assessed on the board, and the most important ideas are selected.

   Tips for moderators for brainstorming or card surveys:
- Written questions must be visible to all workshop participants throughout the brainstorming/card survey.
- Read questions out loud and ask whether the participants have understood them.
- Don't anticipate answers.

- Don't specify/limit the number of Post-it® notes that should be submitted.
- Allow one Post-it® note per idea, maximum of three words per card.
- The writing should be visible from a distance of six to eight meters.
- For card surveys, where anonymity is important, mix the cards thoroughly.

6.1.2 Description of Ideas

There may be plenty of ideas for improvement, and each idea could become a project. However, to spend time and energy efficiently, the relevant ideas should be analyzed in greater detail according to their benefits and drawbacks. A description of an idea is shown as an example in Figure 6.2.

The potential aspects must be broken down into transparent value drivers and/or volume drivers to perform the assessments more soundly, which can be done as shown in Figure 6.3. An example based on this approach is shown in Figure 6.4. Finally, all the relevant ideas for each problem or its individual leverages are summarized in Table 6.1.

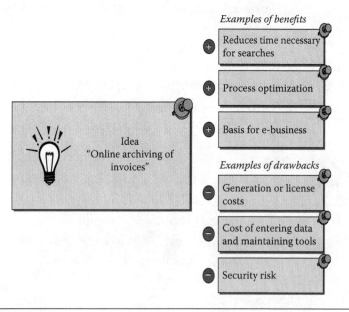

Figure 6.2   Description of an idea—example case.

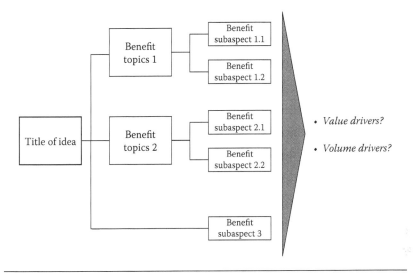

Figure 6.3   Determining potential aspects of an idea.

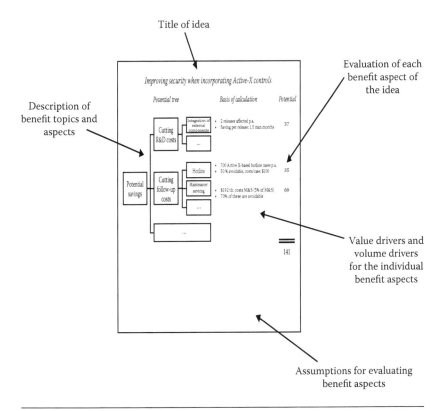

Figure 6.4   Determining potential aspects of an idea—example.

Table 6.1 Ideas Summary Worksheet—Example Case

PROBLEM *Invoice System*

IDEA	DESCRIPTION	REQUIREMENTS	IMPLEMENTATION COST	POTENTIAL
A	Do nothing—Leave system as it is, problems not solved, problems will increase as more suppliers convert to daily deliveries.	No outlay; continue to live with problems	L	L
B	Add personnel to handle increased workload	Two added Accounts Payable personnel	L	L
C	Invoice monthly direct to payables; compare data to packing slips.	Accumulate packing slips, once monthly heavy OT	L	L
D	Have system pay invoices based on receiving data, monthly checks with audit.	Revisions to materials system, some OT for monthly audit	H	L
E	Pay supplier based on shipping report; deliver directly to assembly on multipurpose unit racks.	Some changes to accounting system, eliminate packing materials, change packaging specs, modify panel racks	M	H

IMPLEMENTATION COST: High (H), Medium (M), Low (L)

NOTES: See cost analysis # 121193.

POTENTIAL: High (H), Medium (M), Low (L)

NO.	POTENTIAL ASPECTS	MEASURE (VALUE AND/OR VOLUME DRIVERS)	CURRENT	TARGET	A	B	C	D	E
1	Processing errors reduced	Errors/month	2	0	2	1	1	0	0
2	Accts. Payable overtime reduced	Ave. hours/week	2	0	2	0	2	1	0
3	Morale improved	Transfer requests	2	0	2	1	2	1	1
4	System streamlined	Minutes/invoice	4	1	4	4	80	1	0
5	Supplier relations improved	Complaints/month	1	0	1	0	1	0	0
6	Cycle time to process reduced	Days to process	2	1	2	2	2	1	0
7	Payments simplified	Checks/week	5	1	5	5	1/4	1/4	1/2

6.2 Operational Improvement Techniques

The feasible solutions should be used to utilize the leverages designated for a permanent solution to the problem and to fulfill the objective of the business. In this step, operational improvement techniques
intended for leverages and thus solutions are described. There is an
enormous variety of operational improvement techniques, listed in
Figure 6.5:

- Layout Planning (Systematic Layout Planning [SLP])
- Materials Handling Planning (Systematic Handling Analysis
  [SHA])
- Container Planning (Systematic Container Planning [SCP])
- Planning and Process Improvements (MAXIT)
- Cell Design (Systematic Planning of Cells [SPC])
- Group Technology
- Line Balancing
- Just-in-Time (JIT) Production
- Single-minute exchange of die (SMED)
- Kanban
- Poka-Yoke
- 5S

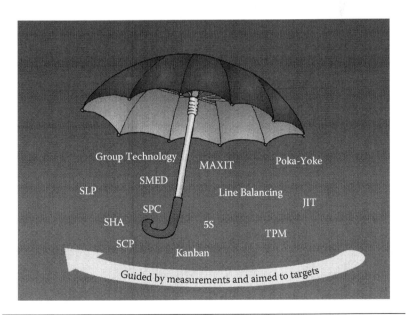

Figure 6.5 Operational improvement techniques.

- Total Productive Maintenance (TPM)
- Kaizen
- ...

6.2.1 Layout Planning—Systematic Layout Planning

The main goals are to reduce material handling and to get more volume from existing floor space. An example layout plan and implementation results of SLP are illustrated in Figure 6.6. Every production or industrial service facility should, wherever possible, be arranged to meet the following considerations (see more in Muther 1974):

- Directness of material flow—back tracking and cross traffic
- Direct labor hours required and utilized
- Investment in material-in-process necessary by the nature of the layout
- Maintenance—space for and ease of; availability of service
- Costs of material handling—by direct labor, handlers, and servicemen—including scrap, packing, returns, salvage, trimmings
- Costs of storing—holding inventory, handling in storage area, controlled areas, ease of identification
- Quality costs—inspection cost, damage to material, access of inspectors and/or test equipment
- Space utilization—idle or wasted space
- Equipment utilization—idle or inaccessible equipment
- Supervision—ease or difficulty to see area, to get operators, to check quality and performance
- Personnel—safety and satisfaction thereof
- Work accountability—planning, scheduling, paperwork, count, timekeeping
- Housekeeping possibilities by nature of layout—including effect on quality, output, equipment and personnel of dirt, dust, fumes, vibrations
- Flexibility—ease of expansion or adaptability to changes in product, process, routing, or schedules
- Setup and tool service—availability of and access for

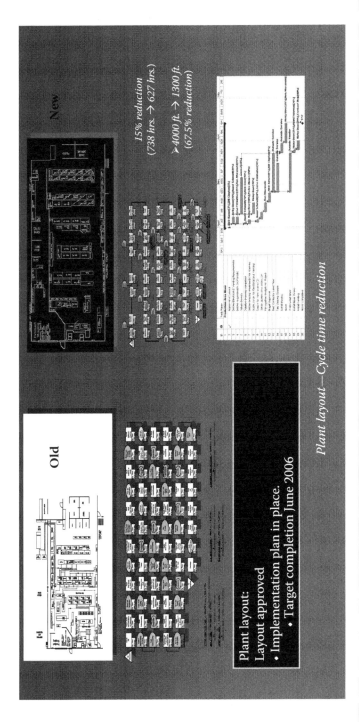

Figure 6.6 Implementation results of SLP—example.

6.2.2 Materials Handling Planning—Systematic Handling Analysis

The goal is to document plan to capture material handling cost savings from the improved layout. An example quantified materials flow diagram on an improved layout, using SHA methodology, is illustrated in Figure 6.7. Every transport and handling of material should, wherever possible, move material to meet ten considerations (see more in Muther and Haganas 1969):

1. Move material toward completion—without back tracking or counter flow.
2. Move material on the same device—without transfers.
3. Move material smoothly and quickly—without confusion or delays, unnecessary handling and awkward positioning or placing.
4. Move material over the shortest distance—without long trips.
5. Move material easily—without rehandling or extra motions.
6. Move material safely—without damage to people, materials, or equipment.

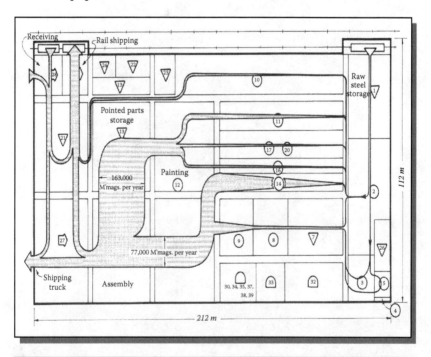

Figure 6.7 Quantified materials flow diagram—example. *Source*: Muther, R., and Haganas, K. *Systematic handling analysis*. Management and Industrial Research Publications, Kansas City, MO, 1969.

7. Move material conveniently—without undue physical effort.
8. Move material economically—without breaking bulk units or making several trips where one would do; combining many small units into one large one.
9. Move material to coordinate with production—without causing production workers extra time and effort by hand handling, bending, or reaching.
10. Move material to coordinate with other handling—without a large amount of different handling equipment that cannot be integrated.

A typical output of SHA should be as shown in Figure 6.8.

6.2.3 Container Planning—Systematic Container Planning

The goal is to reduce material handling by direct and indirect labor. To accomplish this, you need to (see more in Muther and Haganas 1969)

1. Review problems associated with current containers.
2. Identify desired improvements and container features with respect to
   • Ergonomics at points of loading and unloading and kitting
   • Visibility of contents
   • Accessibility
   • Protection
   • Cleanliness
   • Storage density
3. Identify alternatives to current containers.

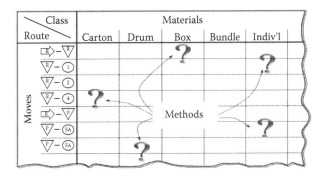

Figure 6.8   What is material handling? *Source*: Muther, R., and Haganas, K. *Systematic handling analysis*. Management and Industrial Research Publications, Kansas City, MO, 1969.

Figure 6.9   Containers—example.

4. Estimate costs and benefits.
5. Evaluate and select the best.
6. Publish a container plan identifying the preferred container for each type of part.

Several container types are illustrated in Figure 6.9.

6.2.4 Planning and Process Improvements

The main goals are to improve coupling of sequential operations, recover associated storage space for value-adding activities, and reduce material handling. To accomplish this, you need to (see more in Plenert 2011)

1. Prepare a detailed multifunction process chart of the current process.
2. Agree on objectives, performance goals, and characteristics of the desired process, including customer service, utilization, setups, and material handling.

3. Review current procedures, rules, and assumptions.
4. Challenge the current process and identify desired improvements.
5. Identify and agree on which classes of parts could be made using a *pull system* and which must remain as *push and expedite*.
6. Prepare a detailed multifunction process chart of the improved process.
7. Identify necessary, enabling improvements in
   - People: organization, roles, responsibilities, job design, and assignments
   - Policies and procedures: rules, timing, guidelines for lot-sizing
   - Information systems

6.2.5 *Cell Design—Systematic Planning of Cells*

The goal is to improve direct labor productivity and space utilization. To do this, SPC (see more in Hales and Andersen 2002) identifies opportunities to more closely couple operations, create group-of-parts, and to make arrangements for reduced travel and material handling, subject to the desired utilization.

The principal physical change made with a manufacturing cell is to reduce the distance between operations. In turn, this reduces material handling, cycle times, inventory, quality problems, and space requirements. Plants installing cells consistently report the following benefits when compared to process-oriented layouts and organizations:

- Reduced materials handling—67% to 90% reductions in distance traveled are not uncommon, since operations are adjacent within a dedicated area.
- Reduced inventory in process—50% to 90% reductions are common, since material is not waiting ahead of distant processing operations. Also, within the cell, smaller lots or single-piece flow is used, further reducing the amount of material in process.
- Shorter time in production—from days to hours or minutes, since parts and products can flow quickly between adjacent operations.

In addition to these primary, quantifiable benefits, companies using cells also report

- Easier production control
- Greater operator productivity
- Quicker action on quality problems
- More effective training
- Better utilization of personnel
- Better handling of engineering changes

A manufacturing cell consists of two or more operations, workstations, or machines dedicated to processing one or a limited number of parts or products. A cell has a defined working area and is scheduled, managed, and measured as a single unit of production facilities.

A cell is essentially a production line (or layout by product) for a group or family of similar items. It is an alternative to layout and organization by process, in which materials typically move through successive departments of similar processes or operations. This layout by process generally leads to higher inventories as parts wait between departmental operations, especially if larger batches or lots are produced. More material handling is required to move between departments, and overall processing time is longer. Exposure to quality problems is greater, since more time may pass and more nonconforming parts may be produced before the downstream department notices a problem.

Three aspects—physical, procedural, and personal—must be addressed when planning a manufacturing cell, which is illustrated in Figure 6.10. Cells consist of physical facilities such as layout, material handling, machinery, and utilities. Cells also require operating procedures for quality, engineering, materials management, maintenance, and accounting. And because cells employ personnel in various jobs and capacities, they also require policies, organization, leadership, and training.

A cell plan is a coupling of parts and process into an effective arrangement and operating plan. It should include

- The layout of operating equipment (physical)
- The method(s) of moving or handling parts and materials (physical)

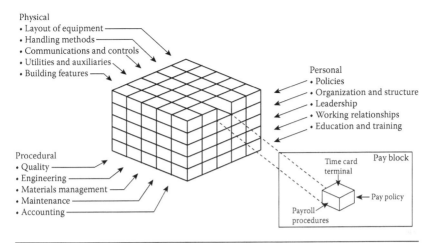

Figure 6.10   Three aspects of a cell design. *Source:* Hales, H. L., and Andersen, B., *Systematic planning of manufacturing cells.* Society of Manufacturing Engineers, Dearborn, MI, 2002.

- The procedures or methods of scheduling operating and supporting the cell (procedural)
- The policies, organizational structure, and training required to make the cell work (personal)

Once the machinery and workplace layout is visualized, the material handling and any storage methods are determined. Material handling equipment, containers and storage, or parts-feeding equipment is added to the layout. The planner also adds any support equipment not already visualized in the workplaces, such as tool and die storage, fixture storage, gage tables and tool setup, inspection areas, supply storage, trash bins, desks, computer terminals and printers, display boards, meeting areas, and so forth.

Once the layout and handling methods—the physical aspects of cell planning—have been determined, the planning team turns its attention to the procedural and personal aspects. The procedural and personal aspects are often more important than the layout and handling in ensuring a successful manufacturing cell. These aspects include the procedures and policies for staffing, scheduling, maintenance, quality, training, production reporting, performance measurement, and compensation. In practice, some of these will have already been determined during the layout and handling discussion and the remainder should be clearly defined by the team.

Cells take different forms based on the characteristics of the parts (P) and quantities (Q) produced, and the nature of the process sequence or routing (R) employed.

Cells are typically used to serve the broad middle range of a *Product–Quantity* (P–Q) distribution. Very high quantities of a part or product—typically more than one million units per year—lend themselves to dedicated mass-production techniques such as high-speed automation, progressive assembly lines, or transfer machines. At the other extreme, very low quantities and intermittent production are insufficient to justify the dedicated resources of a cell. Items at this end of the P–Q curve are best produced in a general-purpose job shop. In between these quantity extremes are the many items, parts, or products that may be grouped or combined in some way to justify the formation of one or more manufacturing cells.

Within the middle range, a production line cell may be dedicated to one or few high-volume items. This type of cell will have many of the attributes of a traditional progressive line, but is usually less mechanized or automated.

Medium and lower production quantities are typically manufactured in group technology or group-of-parts cells. These are the most common types of cells, shown in Figure 6.11. They exhibit progressive flow, but the variety of parts and routings works against a production line.

If the processing steps are specialized in some way, requiring special machinery and utilities, or special enclosures of some kind, then a functional cell may be appropriate.

Functional cells are often used for painting, plating, heat treating, specialized cleaning, and similar batch or environmentally sensitive operations. If the functional cell processes parts for other group-of-parts or production line cells, it will introduce extra handling, cycle time, and inventory because parts must be transported and held ahead of and behind the functional cell. For this reason, planners should examine the practicality of decentralizing or duplicating the specialized process(es) into group-of-parts or production line cells.

The steps required to plan a manufacturing cell are the same for all three types of cells—production line, group technology, and functional. However, the emphasis and specific techniques used will vary somewhat based on the physical nature of the manufacturing processes

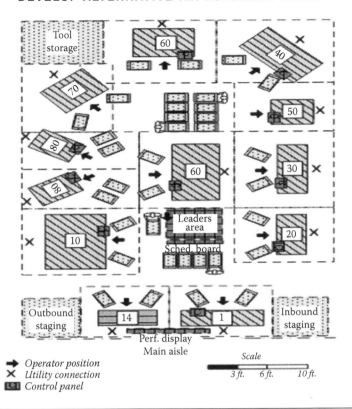

Figure 6.11   A typical group-of-parts cell—example. *Source*: Hales, H. L., and Andersen, B., *Systematic planning of manufacturing cells*. Society of Manufacturing Engineers, Dearborn, MI, 2002.

involved. For example, when planning for machining and fabrication cells, the capacity of key machines is critical and may be relatively fixed. The time required to change from one part or item to another is also critical. Allowances for setup and capacity losses to changeovers are very important. Manpower planning may be of secondary importance, after the number of machines has been determined. In contrast, when planning for progressive assembly, the variability of operation times must be understood, and the work must be balanced among the operators to ensure good utilization of labor. In such assembly cells, utilization of equipment may be a secondary issue.

6.2.6 Group Technology

Classifying the parts simplifies the analysis and design of the cell. Therefore, group technology (see more in Ham 2013) is for the

Figure 6.12   Group of parts—example.

purpose of classification of parts, usually according to their physical characteristics, which is illustrated in Figure 6.12. These include

- Basic material type
- Quality level, tolerance, or finish
- Size
- Weight or density
- Shape
- Risk of damage

Additional common considerations for classification include

- Quantity or volume of demand
- Routing or process sequence
- Service or utility requirements (related to the process equipment required)
- Timing (may be demand-related, e.g., seasonality; schedule-related peaks; shift-related; or possibly related to processing time if some parts have very long or short processing times)

All of these factors can be tied together into a worksheet like the one shown in Figure 6.13. The planner identifies and records the physical characteristics and other considerations for each part or item. If it seems awkward to record or rate each amount or specific dimension, one can rate the importance or significance of each characteristic as to its contrast or dissimilarity with the other parts. Use

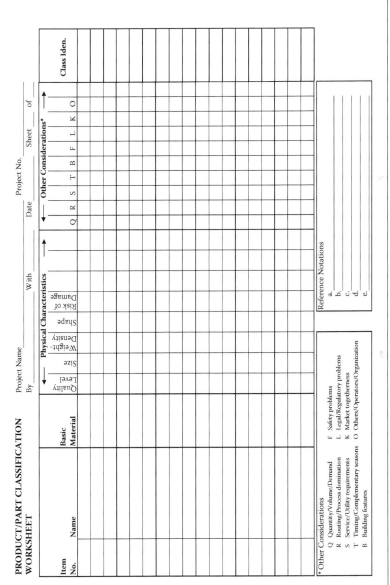

Figure 6.13 Parts classification worksheet. Source: Hales, H. L., and Andersen, B. *Systematic planning of manufacturing cells*, Society of Manufacturing Engineers, Dearborn, MI, 2002.

the vowel-letter, order-of-magnitude rating code shown in Figure 6.13 and defined as

A—Abnormally great
E—Especially significant
I —Important
O—Ordinary

After recording or rating the physical characteristics and other considerations for each part or item, note those parts that have similar characteristics—that is, classify the parts according to the most important characteristics and considerations. Assign a class code letter to each class, group, or combination of meaningful similarities. Enter the appropriate class letter code for each part or item in the Class Identification column.

When a large number of different parts are to be produced, planners should place special emphasis on sorting the parts into groups or subgroups with similar operational sequences or routings. Those assigned to a class will all go through the same operations.

6.2.7 Line Balancing

The planner uses charts and diagrams to visualize the routings for each class or subgroup of parts, and then calculates the numbers of machines and/or operators and workplaces that will be required to satisfy the target production rates and quantities.

In machining or fabrication where the throughput is paced more by the operators than the machines, the planner may need to conduct a line balancing, in addition to capacity and utilization analyses. Also, if the cell performs assembly—such as that of the sheet metal oven— line balancing needs to be conducted.

The best way to visualize the process is with an operation process chart like that shown in Figure 6.14. In addition to showing the progressive assembly of the finished item, the process chart also shows the labor time at each step. Given a target production rate and the number of working hours available, the planner calculates the work content of the process and breaks it into meaningful work assignments.

In this way, the required number of operators and workplaces is determined, along with the flow of materials between them.

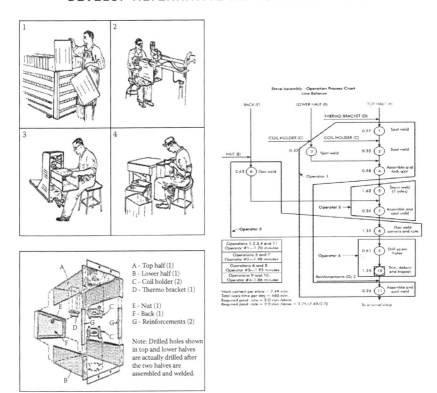

Figure 6.14 Line balancing using operation process chart—example. Stove assembly—Operation process chart line balance. *Source*: Hales, H. L., and Andersen, B., *Systematic planning of manufacturing cells*. Society of Manufacturing Engineers, Dearborn, MI, 2002.

Assumptions or calculations must be made to establish the time that will be lost to breaks and to *non-value-adding* tasks such as material handling, housekeeping, and the like. The formal name for this process is line balancing (see more in Scholl 1999).

A good line balance achieves the desired production rate with the minimum number of operators and minimal idle time.

6.2.8 Just-in-Time Production

JIT is a technique that is used for producing the required customer order within the required amount of time by using minimum equipment, labor, and space resources (see more in Monden 2011).

JIT is established on principles such as ensuring flow-type production, deploying takt-time implementation, equalizing production tempo to demand rate, and establishing a pull-type production system.

It creates a system that does not tolerate disruptions in the production system, minimizes waste in production, and reduces flow time. The following are characteristics of JIT production systems:

- Equipment is placed with respect to the operations' sequences
- Small and cheap devices
- Single-part production flow
- Multiskilled workforce
- Easily activated/deactivated operations
- U-type manufacturing cell layouts
- Set production tempo with respect to takt time
- Defined standard operations

6.2.9 Single-Minute Exchange of Die

Model Turnaround Time is the period elapsed between the production of the final part of a lot and the production of the first flawless part of the following lot. Model Turnaround consists of the elements of replacement; placement; and adjustment of parts, tools, and equipment (see more in Shingo 1996).

Model Turnaround is a production technique with a major contribution in fulfillment of JIT production, ensuring fulfillment of model adjustment within a minimum time. Model Turnaround can be shortened by shorter Model Turnaround, more frequent model turnarounds, smaller lot sizes, shorter transition periods, less in-process inventories, and higher competition power.

Model Turnaround activities consist of the activities (inner activities) required to be performed by stopping the machine (without producing a part) and the activities to be executed (outer activities) while the machine is in operable status (while performing production).

Model Turnaround shortening steps consist of disintegration of inner and outer activities, transforming inner activities into outer activities, and shortening all activities. Model Turnaround generates a list of required equipment for model turnaround in the stage of disintegration of inner and outer activities; it inspects whether all equipment is running and kept in suitable condition; and it ensures all equipment is kept available at the worksite.

It is necessary to ensure that activity requirements prior to model turnaround, functional standardization, and intermediate devices are used in the stage of transforming inner activities into outer activities.

Within the stage of shortening of all activities, it is necessary to ensure that parallel activities and retaining mechanisms are developed, adjustments are prevented, and automation of activities is provided. Model Turnaround improvement activities start and end with 5S. Adjustments after model turnaround should be minimized; mold standardization used on the workstation should be applied; and all installation activities should be standardized.

6.2.10 Kanban

The purpose of Kanban, which also means *card*, is to help the realize JIT production (see more in Monden 2011).

In the *pull type*, the following operation takes the required parts from the previous operation at any moment and in the amount it needs. Similarly, the previous operation produces the pulled amount of the next operation. Kanban is a mechanism that is used to run this system, which is illustrated in Figure 6.15.

6.2.11 Error Prevention (Poka-Yoke)

While Poka means carelessness and thoughtlessness, Yoke means eliminating them. Poka-Yoke aims to reach zero-defective production by using a myriad of error-preventive and ancillary tools and

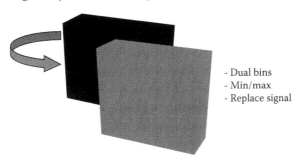

- Dual bins
- Min/max
- Replace signal

Figure 6.15 Illustration of Kanban system.

approaches (but without the need for more control elements) against the cases arising from human factors such as forgetfulness, inattention, misunderstanding, lack of concentration, deficiency of standards, lack of experience, neglect, sabotage, and so forth. For this purpose and if needed, additional mechanisms may be added to the workstation and/or the design of the product may be altered.

One of the main causes of a problem is human error, which can occur for many reasons:

- *Forgetfulness*: Employees fail to focus on the work performed most of time; for example, an employee may forget to tighten a nut on any workpiece with required torque.
- *Misunderstanding*: A quick leap may produce an incorrect result without thorough observation of the situation in question. For example, if meetings are often held in room 104, a meeting call that states room 401 may be read as 104.
- *Reading errors*: An employee may incorrectly read, in cases such as quick reading or remote distance reading, a code number on a label or a number on the indicator. For example, the number 10001 may be read as 100001.
- *Lack of experience*: Errors can occur where sufficient on-the-job training was not offered, especially for new employees.
- *Short circuit*: Errors can arise from the assumption that the result will not change even if we do not apply the rules in some cases. For example, a person may run a red light if there is no vehicle in sight.
- *Lack of awareness*: An employee may make an error involuntarily while daydreaming.
- *Absence of a standard*: Errors can occur if work instructions relating to procedures are not available and the things to be done when necessary are left to the employee's initiative.
- *Surprise errors*: Errors can arise from improper operation without a visible cause related to the equipment or system.
- *Errors resulting from slowness*: Errors caused by delays in decision making; for example, a person taking too long to press the brake pedal when learning to drive a new vehicle.
- *Deliberate*: Though very rare, this kind of error can still be seen. It is done deliberately to damage the business.

Poka-Yoke integrates tools with production or assembly equipment to prevent an error from turning into a defect, so as to fulfill the control function by taking into account that humans may inevitably make a mistake.

In Poka-Yoke implementations, the following steps are recommended (see more in Shingo 1986):

1. Data collection and analysis
2. Prioritization
3. Determination of alternative Poka-Yoke techniques
4. Selection of optimum implementation
5. Tracking
6. Standardization and dissemination

Poka-Yoke elements comprise the limiting switches, illuminated warnings, templates, guides, sensors, pressure switches, setting pins, counters, and so forth. The main functions are shutdown/stop, control, and warning.

During Poka-Yoke applications, suitable Poka-Yoke tools are selected in a way that they will identify deviations from the standards set according to the characteristic features of the product. Mechanisms that determine the deviations from procedures and parameters are established, a typical example of which is illustrated in Figure 6.16.

Figure 6.16   Typical Poka-Yoke example.

Poka-Yoke systems are based on three main techniques:

- Contact techniques
- Constant value techniques
- Motion techniques

Each of these techniques can be matched with a *control* or a *warning* system. In addition, each of these techniques incorporates Poka-Yoke designs and applications using a different approach and prevention of defects at its source.

In the design of process-specific Poka-Yoke systems, the decision on which technique is more suitable for preventing the occurrence of a defective product in terms of your process is made by considering the features of the part produced part and process above all.

- *Contact techniques*: This is established on the principle of whether or not the contact technique causes formation of *physical* or *energy contact* with a sensor of a product or component. For instance, if a limit switch is in a position in which a screw is absent in the transition of the installation conveyor and a piece lacking a screw installation is pressed, the acoustic warning circuit remains in the off position and a siren sound warns the operator, as illustrated in Figure 6.17.

  The same Poka-Yoke application can also be accomplished with the energy contact. Since there is no obstacle in front of a

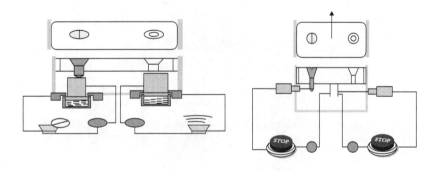

Figure 6.17   Poka-Yoke touch technique—example.

photoelectric light beam dispatched from the right-hand side with the missing screw, this completes the circuit by contacting the detector and electrically causing a pressing of the stop button, ensuring that the assembly conveyor is disconnected, as illustrated in Figure 6.17.

• *Constant value techniques*: Constant value techniques are Poka-Yoke implementations used where a specific number of components are to be installed on a product or an operation is repeated a specific number of times on a workstation.

Constant value techniques are generally equipped with limit switches dispatching a signal to a counter in each motion. When the designated number of signals have been completed, the counter activates a signal mechanism or releases the piece through a valve.

For example, in 10-spot welding, prior to Poka-Yoke, the operator may have bypassed one or more welding points. After improvement, a control unit dispatches on and off commands to a pneumatic valve if the number of presses is 10 and a counter counts the number of button presses for each welding procedure. Otherwise, the operator is prevented from taking the piece from a fixture and sending it to the next operation.

• *Motion techniques*: This technical process is based on the principle of control—whether a motion or operation has been completed within a designated time (e.g., operation cycle time) or not. For instance, a specific time has been identified for affixing a tag to the part. If a photoelectrical sensor recognizes the tag for a period longer than the designated one, the operation is blocked (line is stopped) because an identification tag had not been affixed to the previous part. This can also be used in detecting whether or not these technical operations have been fulfilled within a consecutive order.

Sensors and devices that are used in Poka-Yoke implementations are grouped under three main categories:
• Physical contact sensors
• Energy sensors
• Physical change sensors

Each of these three categories contains a wide range of sensors and detection equipment suitable for use in different cases. Some Poka-Yoke examples include

- A Poka-Yoke application integrated with a shower warning system changing color according to water temperature, as illustrated in Figure 6.18
  Error: Taking a shower while the water is too hot/cold
  Defect: Production of burns or chills
- A Poka-Yoke application integrated with a control system for preventing overflow of water from the sink. If the tap is left on, the water running over a specific level is discharged from the upper hole, as illustrated in Figure 6.19.
  Error: Forgetting that the tap is on
  Defect: A flooded house

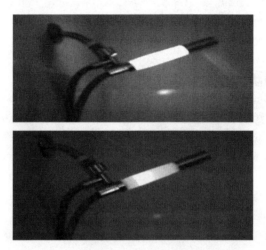

Figure 6.18   Example Poka-Yoke implementation.

Figure 6.19   Example Poka-Yoke implementation.

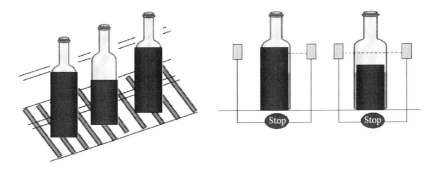

Figure 6.20   Example Poka-Yoke implementation.

- Drinking bottles in a filling facility are placed in cardboard boxes at the automatic filling machine in order to be shipped. Due to clogging occurring at the filling workstation, some bottles are placed in cardboard boxes without complete filling and may be delivered in this way to the customer, as illustrated in Figure 6.20.

As an improvement, an *electron beam sensor* should be placed to identify the filling level of the bottle. If the bottle is not at the required filling level, an electron beam reaches the detector and triggers the stop button, enabling the conveyor belt to stop, as illustrated in Figure 6.20.

6.2.12 5S

Prior to commencing a systematic or focused improvement in a business, applying 5S should be an ideal option to ensure clean-up, arrangement, and order.

Though 5S application is a technique requiring very simple understanding and very low amount of expenditures, once applied meticulously it ensures macro-level improvements without engaging specific lean production techniques in the areas of occupational safety, productivity, quality, and equipment failures (see more in Liker 2004).

The purpose of 5S is to reach a flawless point in waste, work accident, failure, scrap, installation, delay, and complaints:

Seiri	(Sorting)
Seiton	(Simplifying)
Seiso	(Sweeping)
Seiketsu	(Standardizing)
Shitsuke	(Self-discipline)

Before starting 5S-related activities, the first thing that should be done is to record videos and take pictures of the application area. During the progress of 5S applications, these photos will be very helpful in terms of benchmarking, as shown in Figure 6.21.

Sorting: Sort out the necessary from the unnecessary and discard the unnecessary: work-in-progress, unnecessary tools, unused machinery, parts, information, defective products, papers, and documents. Figure 6.22 shows how to sort.

Sorting—placing goods in the right places, categorizing goods, finding the real reasons for dirt, getting rid of difficult areas to be cleaned, getting rid of unnecessary ones, finding the causes of dirt and leaks, cleaning floors—uses methods such as sorting warehouses.

Simplifying: Simplifying is creating a designated and marked place for everything according to the frequency of use, as illustrated in

Figure 6.21   Before 5S—example.

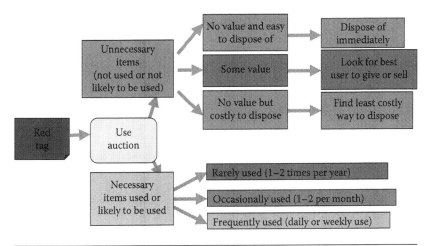

Figure 6.22   5S: Sorting.

Figure 6.23. The goal is achieved when the items used most often are within easy access, thus reducing the time to find something to almost zero. Even someone who does not work in the area could put everything back where it belongs by the way objects are marked. Simplifying steps:

1. Review all frequently used items and determine where to put them. Put those used most often closest to the work area and those used less often farther away.

Figure 6.23   5S: Simplifying.

2. Develop a way to label or show where everything goes. Consider
   - Shadow board
   - Marking the item and the location
   - Color-coding
   - Labels on drawers with a list of contents inside
3. Develop ways to replace usable items daily:
   - Establish lead times for replacement of daily usage supplies.
   - Determine minimum and maximum supply levels and mark them.

Simplifying uses techniques/approaches such as identifying functional layout plans and defining, classifying, and signing the areas, thus making it possible to access the required items when needed swiftly; minimizing the search effort; applying the rule of first in, first out; and placing open warning panels. Figures 6.24 and 6.25 shows after examples of simplifying.

Sweeping: Physical and visual control of the work area. Sweeping is done when regular sweeping processes occur and areas are clean, safe, and neat.

Sweeping uses approaches such as disposing of garbage, trash, and foreign materials; always being clean as if you are ready for inspection

Figure 6.24   5S: After simplifying—example.

Figure 6.25   5S: After simplifying—example.

at all times; identifying individual responsibilities; initiating cleaning campaigns; and conducting cleaning checks. Sweeping actions:

- Determine a regular schedule for cleaning the yard, work, and break areas.
- Orient everyone including new employees to daily 5S activity responsibilities and expectations.
- Post area cleaning guidelines and schedules.
- Keep tools, machinery, and office equipment clean and in good repair.
- Keep the yard, work/break areas, and trailers clean and orderly.
- Establish a dependable, documented method to reduce hazardous waste and minimize usage of chemical products.
- Perform safety inspections on a regular schedule.

Standardizing: Standardizing is creating standard ways to keep the work areas organized, clean, and orderly and documenting agreements made during the 5S's. For standardization to be successful, employees must understand the value of using and maintaining standard methods. How to standardize:

- Use a 5S assignment map to help everyone know exactly what they are responsible for doing, when it is to be done, and where and how it is to be done to maintain the first 3S agreements.
- Have clear instructions for people who deliver goods or materials to the site. Clearly mark and post where the material, tools, and equipment are to be placed. Educate the supplier on what is expected of him or her.
- Use a standard 5S format for communication boards/binders so they are similar in appearance.
- Install standard visual controls for the area (signboards, shadow boards, outlining, etc.).
- Develop standard labeling and outlining methods for the area or department so that anyone can see when something is out of place.
- Document all 5S agreements and implement any changes.

Standardization uses approaches such as identifying ideal status, standard solutions and responsibilities, marking hazardous areas, using tags, performing functional marking, using functional color indicators, arranging cables, marking upper and lower limits of control points and sensitive maintenance points, providing transparency, and maintaining order and continuity of the organization.

Self-discipline: Providing training and discipline, performing accurately at all times in accordance with the rules, creating good habits, creating a disciplined work area, routinely cleaning the environment, using safety clothing, managing public areas, and scheduling emergency drills are some examples of self-discipline. Self-discipline is evident when

- The 5S rules for sorting, simplifying, sweeping, and standardizing are being followed.
- All changes have been documented.
- A daily 5S activity checklist is posted and used.
- The 5S communication board/binder is updated regularly by personnel listed as responsible.
- The work area is kept neat and clean.

No matter how successfully 5S is implemented within the business, there will be new participations after a specific period of time and 5S activities can be extended into other areas besides the model area.

5S scores obtained as a result of audits conducted by top management using a monthly 5S Control List in the 5S application area are a criterion measuring achievement of implementation.

After the audits conducted for both 5S steps and certification, the auditor must complete a Corrective Action Form at all times.

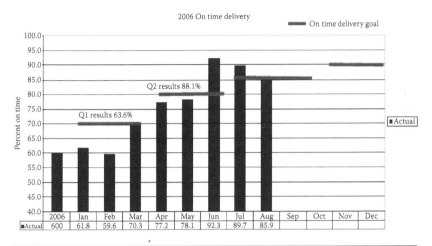

	2006	Jan	Feb	Mar	Apr	May	Jun	Jul	Aug	Sep	Oct	Nov	Dec
Actual	600	61.8	59.6	70.3	77.2	78.1	92.3	89.7	85.9				

Figure 6.26 5S: Benefits—example.

Figure 6.27 After 5S—example.

This form is also intended as a training tool for the 5S team and should be presented on the Activity Board and converted to action plans.

As mentioned at the beginning, the cleaning and order provided by 5S implementation should be evaluated with respect to

- Increase in productivity
- Decrease in work accidents
- Decrease in equipment breakdowns
- Decrease in defective production.

Figures 6.26 and 6.27 show different benefits of using 5S.

6.2.13 Total Productive Maintenance

Maintenance systems have passed through breakdown, preventive, predictive, and productive maintenance stages and today reached the total productive maintenance stage. The goals of total productive maintenance are to extend the life of equipment, to keep factories and equipment in optimum condition for production and/or services, to increase the return on investment, to increase the ability to cope with emergencies, and to provide safety and security (see more in Liker 2004).

This is carried out by the total participation of all employees and all units, from management to operator. Through the support of total productivity maintenance, global facility productivity is maximized by enhancing the productivity of the total equipment; the required maintenance systems of machinery/equipment are established for their entire life; and process scrap rates, workstation and line stops, workstation failures, and accidents are diminished.

Among the losses that total productivity maintenance deals with are stop losses (failures, installations and setups, cutting tool and jig replacement), equipment performance losses (idle times and small stops, low tempo), scrap losses (scraps and reprocessing losses), closing losses, workforce losses in production (management losses, motion losses), organization losses, logistics losses, measurement and adjustment losses, material losses, energy losses, and so forth. Total productivity maintenance consists of

- Preparation stage (announcements of senior management decisions, preparation of training and campaigns, preparation of organization, identification of main objectives, preparation of a project plan)
- Application stage (establishment of systems to improve the efficiency of equipment, development of autonomous maintenance program, deployment of training for the development of operations and maintenance qualifications, development of an equipment management program)
- Continuity stage (disseminating and upgrading the level)

Since TPM is a technique intended to reduce the losses arising from the lower level of Overall Equipment Efficiency (OEE), the business should have a collection system for loss data transformed into monetary losses in relation to all workstations and facilities of the business.

$$OEE = Availability \times Performance \times Quality\ Factor$$

$$OEE = \frac{Working\ Time}{Loading\ Time} \times \frac{Net\ Working\ Time}{Working\ Time} \times \frac{Productive\ Working\ Time}{Net\ Working\ Time}$$

After the establishment of the monitoring system of the key performance criteria such as OEE, Mean Time between Failures (MTBF) and Mean Time to Repair (MTTR), the most current approach of TPM implementation called JIPM's eight building block stages will be carried out.

6.2.14 Kaizen

Kaizen is a corrective technique based on a feedback mechanism. By means of a gradual and continuous change, small investments, shared information, and long-term effects, Kaizen results in change for the better. In addition to increasing profitability by solving problems and decreasing losses, Kaizen (see more in Liker 2004) also provides additional gains such as

- Resolving the problems within the business with the participation of all staff including employees

- Delegation (transferring the authority and responsibility to lower levels)
- Ability to assign performance-based business objectives to teams
- Budgeting improvement activities for future periods

In its first stage, Kaizen works to change the current situation for the better. To do this, initially small (individual) and medium (small group) changes are recommended. Then major changes (inventions) follow.

In Kaizen teamwork, the selected team leader and team members are present. Depending on the complexity of the problem and the solution techniques used, the team would consist of three to seven people. Although it varies according to the nature of the problem, the Kaizen team can consist of the following people:

- Team leader (usually area supervisor)
- Member (responsible for the process)
- Member (responsible for the subprocess in which problem occurs)
- Member (responsible for the internal supplier process)
- Member (responsible for the internal customer process)

Kaizen steps

The Kaizen steps comprise eliminating (removing, stopping, excluding), reducing (simplifying, centralizing, synchronizing, integrating, merging), and changing (alternative creating, replacing, transforming, separating). During the implementation of Kaizen, the steps that need to be applied are

1. Plan
    - Definition of the subject
        - Determination of subtitles of the subject
        - Determination of problem
    - Collection of team
    - Preparation for macroplanning
    - Investigation of the problem
        - Data analysis: 5W&1H analysis
        - Physical analysis
    - Investigation of the system

- – Investigation of the system/process
- – Determination of ideal process conditions and parameters
- Determination of challenging goals
  - – Determination of Specific Measurable Achievable Realistic Time-Bounded (SMART) goals
  - – Calculation of potential returns
- Cause analysis

2. Do
  - Listing potential solutions
  - Matrix analysis
  - Enabling risk analysis of potential solutions
  - Implementing solutions that remove root causes

3. Check
  Tracking results

4. Act
  - Managing changes with respect to job safety
  - Communication and training
  - Updating operational procedures
  - Managing documentation changes

Basic rules for implementation:

- Abandon the traditional rigid ideas relating to production.
- Consider how it will be performed rather than why it will not be performed.
- Do not look for perfection. Even if it achieves only 50% of the target, do it.
- Prefer that ten persons are smart rather than one person is knowledgeable.
- Don't come up with fake excuses. Start questioning the current practices.
- Correct the mistakes immediately.
- Creativity happens when in the face of difficulty.
- Ask five times why and try to find the underlying cause.
- Keep in mind that there are infinite opportunities.

Kaizen types

There are four different types of Kaizen that are differentiated based on the appearance, complexity, and root causes of the problem.

Though they all use the Plan–Do–Check–Act (PDCA) cycle, they differ in details, duration, and number of participants and the level of techniques.

- *Quick Kaizen*: This is developed independently by an individual from the generation of improvement ideas to finalization using the resources of the business. The process is concluded in a few days and simple techniques are used.
- *Standard Kaizen*: This is developed independently by an individual from the generation of improvement ideas to finalization using the resources of the business and having the support of persons with expertise. The process is concluded in a week and medium-level techniques are used.
- *Comprehensive Kaizen*: The Kaizen team focuses on improvement using a guide support. The process is concluded in a few months and intermediate techniques are used.
- *Advanced Kaizen*: This is focused improvement work completed with the support of an expert or a guide whose solution to the problem is sought by a Kaizen team. The process is completed in six months and advanced techniques are used.

7

EVALUATE IMPROVEMENT IDEAS AND SELECT THE BEST ONES

Here the improvement ideas most suitable for the business will be selected. To do this, evaluate the alternative improvement ideas developed in the previous step for each corresponding problem or its individual leverages. In this step, potential improvements are evaluated according to their net potential quantitative and qualitative effects.

Also, alternative improvement ideas should be evaluated according to the risk factors to which they may be susceptible with their probability of occurrence. Therefore, both by evaluating according to quantitative and qualitative factors and considering possible risk issues, the best improvement ideas are selected.

Only by making an evaluation as objectively and impartially as possible and by seeking sufficient responses to the questions that follow can we come up with the best decision.

- Which idea(s) are more effective in helping to achieve the target?
- Are selection criteria (sustainability, quality, cost, transition time) appropriate?
- Is the idea sufficiently defined?

If more than one improvement idea is implemented for the same problem, the ideas may affect each other in a negative way or eliminate their effects. In this case, it is also not possible to determine the effects of each individual improvement idea on the target. Therefore, improvement ideas that need to be considered together should be evaluated and selected. Ideas should be chosen that can be executed in harmony with each other, making it easy to create implementation plans.

Why you do it

Now that you have developed alternative courses of action (i.e., improvement ideas), you need to select the best ones for each problem or its individual leverages. That selection should match the expected goals you defined earlier. Once the decisions are made, you need to ensure that they are accepted by the people who ultimately will make them work, and are approved by the people who have responsibility for their impacts.

Deliverables

- Selected improvement ideas for each problem or its individual leverages
- Prioritized selected ideas

Approaches and techniques used

Typically, this selection will be based on comparisons of both quantitative and intangible factors. Table 7.1 is an example form for evaluating the alternatives on the basis of their quantitative factors. Quantitative factors, on their own, are rarely sufficient for selecting the best plan.

Therefore, in addition to the quantitative factors, qualitative ones should be considered as well. The weighted-factor technique is the most effective way to make this type of selection (Muther 2011). After making a list of potential aspects of ideas as specific factors (determined in step 4), weights should be assigned to indicate their relative importance.

Table 7.2 is a typical factors analysis form. Factors analysis can prevent your overlooking an important factor. In addition, it allows the key users and approvers to be part of the decision-making process. Since ratings are subjective, they should be done by several persons. Next, the evaluation team members should rate the performance or effectiveness of each alternative idea prepared in step 4 based on each weighted factor and select the best ones. It is better to have each make their own ratings independent of each other. Then they can get together to compare the results and resolve any differences.

Table 7.1   Evaluation Form by Quantitative Factors—Example Case

	Bus. _____			No. _____
	By _____			Date _____

Ratings by .... Approved by ....

Problem: *Invoice system*

Description of Alternative Improvement Ideas:

- A. *Do nothing*
- B. *Additional person*
- C. *Invoice monthly*
- D. *Monthly audit*
- E. *Acc. sys. change*

	QUANTITATIVE FACTOR		ALTERNATIVE				
			A	B	C	D	E (d)
1.	Annualized Net Savings (Sales Inc./Cost Red.)	c	0	0	0	(...)	250,000
2.	Personnel						
3.	Material						
4.	Capital						
5.	Others						
6.							
7.	Annualized Investment Cost	a	0	0	0	0	33,800
8.		b					
TOT.	Annualized Net Savings over Investment Cost		0	0	0	(...)	216,200

Reference Notes:

a.	Total Capital Investment/Expected Life (yrs.)	d.	Exp. Life 5 yrs.
b.	One-time costs don't need to be considered	e.	
c.	Potential savings over differential implementation costs (yrs.)	f.	

- Identify each alternative. Label the alternatives with letters.
- Establish all pertinent factors, considerations, or objectives affecting the choice of the best ones.
- Assign to each factor a weight value indicating its relative importance to the effectiveness of the improvement ideas. Select the most important factor and assign to it a weight of 10. Select the least and assign it a low number such as 1, 2, or 3. Weigh the importance to the factor relative to the most important (10) and the least important.
- For each factor rate the effectiveness of each alternative in achieving that factor's objective, using A, E, I, and O to

Table 7.2 Factors Analysis Form—Example Case

EQUIVALENT WEIGHT	Bus. _____ No. _____
	By _____ Date _____
Weight set by .... Tally by	**Problem:** *Invoice system*
	Description of Alternative Improvement Ideas:
Ratings by .... Approved by	A. *Do nothing*
EVALUATING DESCRIPTION	B. *Additional person*
A = Almost Perfect, O = Ordinary Result	C. *Invoice monthly*
E = Especially Good, U = Unimportant Result	D. *Monthly audit*
I = Important Result, X = Not Acceptable	E. *Acc. sys. change*

	FACTOR/CONSIDERATION	WT.	ALTERNATIVE									
			A		B		C		D		E	
1.												
2.	Payments simplified	5	U	0	U	0	A	4	A	4	A	4
				0		0		20		20		20
3.	Processing errors reduced	10	U	0	I	2	I	2	E	3	A	4
				0		20		20		30		40
4.	Accounts payable overtime reduced	8	U	0	A	4	U	0	I	2	A	4
				0		32		0		16		32
5.	Morale improved	7	U	0	I	2	U	0	I	2	I	2
				0		14		0		14		14
6.	System streamlined	2	O	2	O	1	U	0	E	3	I	2
				4		2		0		6		4
7.	Supplier relations improved	4	U	0	A	4	U	0	A	4	A	4
				0		16		0		16		16
8.	Cycle time to process reduced	4	U	0	U	0	U	0	I	2	E	3
				0		0		0		8		12
TOT.	Weighted Rate Down Total			4		84		40		110		138

Reference Notes:
a. _____ d. _____
b. _____ e. _____
c. _____ f. _____

Source: Richard Muther, *Planning by design.* Institute for High Performance Planners, Kansas City, 2011.

represent a descending order of effectiveness. Work across the form from side to side rather than from top to bottom in each column. By doing this you can be sure to keep the same meaning for a given factor as you move from idea to idea.

- Convert all letter ratings to numbers (A = 4, E = 3, I = 2, O = 1), and multiply by the previously established weights.
- Total the weighted rate values for each alternative. The highest ones can be indicated as the best available ones.

Typically too many factors may need to be involved. The following factors can be considered in evaluation:

- Cycle time
- Productivity

- Sales/market growth
- Capital reduction
- Implementation duration
- Service/response
- Effectiveness
- Sustainability
- Eliminate duplication of effort
- Saving energy
- Simplicity
- Applicability
- Risks and uncertainty
- Effect on safety and environment
- Flexibility
- Acceptance by key employees
- Effect on working relationships
- Effect on supplier relationships
- Effect on quality

We should start spending our time and energy on the selected ideas with the highest priorities. The Priority Worksheet shown in

Table 7.3 Priority Worksheet—Example Case

	DATE	IDEA—PROBLEM DESCRIPTION	IMPORTANCE	URGENCY	PRIORITY
1	7/13	Assembly team vacation schedule (Simplify for next year)	E	2	6
2	7/13	Setup time—IS shear (Capacity problem)	A	7	28
3	7/13	Maintenance work order flow (Lost slew work orders)	I	5	10
4	7/13	Nonreturnable packaging disposal (Housekeeping in receiving)	I	6	12
5	7/13	Lunch room serving area arrangement (Congestion/confusion)	0	3	3
6	7/13	Redesign bottom plate—Model #1206 (Fit-up problem)	E	9	27
7	7/17	Changes on accounting system (Invoice system problem)	A	9	36

IMPORTANCE: A = Absolutely Necessary, E = Especially Important, I = Important, 0 = Ordinary Importance, U = Unimportant.
URGENCY: Use a 1 to 10 scale; for example, 10 = today, 7 = this week, 4 = this month, 1 = this year.
PRIORITY: Convert Importance vowels to numbers (A = 4, E = 3, etc.) and multiply by Urgency values.

Table 7.3 recognizes that both importance and urgency must be considered. Accordingly the priority of each idea is determined, which is the product of importance and urgency. To rate importance, use vowel letters. These will be relative ratings.

Assignment of an urgency rate to an idea should be done by considering the risks that may be incurred because of delaying the implementation of that idea (i.e., delaying the solution for that corresponding problem). For urgency, use a 1 to 10 scale, where 10 means immediate and 1 means sometime this year.

8

IMPLEMENTATION PLAN OF SELECTED IMPROVEMENTS AND CONTROL

Once the improvement proposals are approved and accepted, they are ready to be installed. In this step, you will develop a schedule, do the installation, and follow up with an audit once the improvement is in place. This step stipulates the results audit to ensure that the benefits and savings are actually realized, as well as to give to management a measure of the value of the improvement efforts.

For each improvement project, we should decide and show where, how, when, and by whom the actions will be realized in the implementation plan. Selected projects are implemented throughout a certain period of time, and in the meantime the data are gathered from the designated measuring points within the determined intervals. Accordingly, deviations from the project targets are evaluated, their causes are analyzed, and necessary adjustments are made.

As a result, if the desired total business target is not reached, according to the developed additional improvement recommendations, changes should be made on the improvement plan. In short, all the effort is directed to implementing new solutions effectively and gaining measurable and sustainable benefits.

Why you do it

Turning the selected best improvement ideas into implementation plans and implementing them to ensure that the potential benefits are actually realized to reach the business financial and improvement targets

Deliverables

- Project plan for recommended improvements
    - Assignment of people responsible for implementation (i.e., who performs the action, and under whose area of responsibility does it fall?)
    - Specifying all important implementation requirements with a detailed timetable
    - Taking account of compatibility with start and end dates of implementation
- Quantified potentials and costs of the implemented actions and their comparisons with targeted goals
- Tasks and responsibilities in implementation and control

8.1 Implementation Plan

After the evaluation of prioritized potential improvements, we reach a step in which the related activities of the ideal solution (project) of each problem or leveraged problem is performed with the help of a detailed implementation plan. Each implementation plan should be directed to a specific project and must not be in conflict with the other projects.

Patterned after the traditional Gantt Chart (see more in Kerzner 2013), an installation schedule provides a place for the actions required, responsibility for each action, and a timeline for accomplishment. The right-hand column allows notations and further action.

Team members should include people who are most likely to be directly affected by the project results and are interested in making improvements. Decide who will be the team leader (or facilitator), who will be full members, and who outside the team may be called upon for support. It's also a good idea to list contact information for everyone. Table 8.1 shows an example of a project team's contact information.

The team should develop a separate project schedule for every project. Figure 8.1 shows a section of a typical installation schedule form. Estimated time requirements for each action are listed in the schedule, as well as who will do what. Determine completion dates for each action and list them.

Table 8.1   Project Team's Contact Information—Example Case

PROJECT TEAM				HRS. REQ'D.	
NAME	ROLE	DEPARTMENT	PHONE	EST	ACT
Rick Saginaw (RS)	L	Receiving	X323	8	
Bill Payer (BP)	M	Accts. Payable	X108	8	
Annie Price (AP)	M	Purchasing	X222	8	
Mac Dawes (MD)	M	Data Processing	X289	8	
Terry Martin (TM)	S	Industrial Eng.	X579	3	
Bea N. Kountre (BK)	S	Controller	X101	1	
Link Missing (LM)	O	Supplier	555-1121	4	
			TOTAL	40	

Note: L = leader, M = member, S = support, O = other.

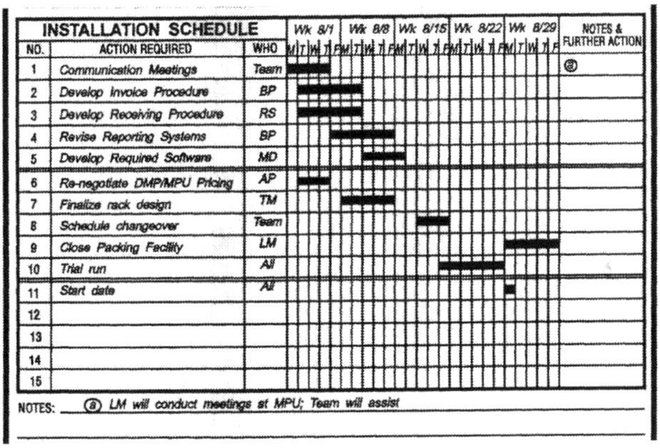

Figure 8.1   Installation schedule—example case.

8.2 Monitoring and Evaluation

Improvement planning systematics have a dynamic and continuous character. During the improvement planning process, feedback is provided using the information obtained as a result of control (monitoring and evaluation) activities. Monitoring is the regular follow-up and reporting of progress toward goals specified in the improvement projects. Evaluation is the measurement of implementation results against goals and analysis of the consistency and relevance of these goals.

Review of improvement projects involves comparison of the results targeted and attained. Project realizations are reviewed in terms of

timing and relevance for goals. As a result of this review and evaluation, every selected project (improvement idea) is confirmed and
implementation continues if

- Implementation plans and actions are implemented as planned.
- Progress toward achievement of goals is in line with expectations.

On the other hand, if changes are observed in terms of the aforementioned considerations, unexpected or undesired results occur, or
existing goals are not realistic, then the project is revised, reevaluated,
and continues to be implemented with its updated version.

Effective implementation of monitoring and evaluation actions
requires linking the goals established in the improvement projects to
objective and measurable performances before proceeding into implementation, which is illustrated in Figure 8.2.

Monitoring action consists of the regular reporting of progress
toward achievement of goals and its submission to related parties.
Reports focus mainly on actions, resources, and outputs. The contents
of reports, the frequency with which they will be prepared, the units
by which they will be prepared, and the authorities to whom they will
be presented have to be identified. A monitoring report must contain
the following elements:

- Projects
- Actions and goals (time, potential, cost, quality)
- Explanations and comments about realizations
- Information about the current status

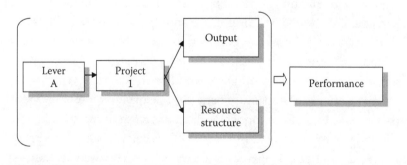

Figure 8.2 Leveraged problem, project, output, and performance relationships.

Besides enabling management to evaluate goals, the monitoring report would help the business take measures rapidly and effectively against unexpected cases. In addition to areas where progress has been achieved, the issues on which progress could not be achieved must also be reported.

Evaluation is a comparative analysis of implementation results and predefined goals. This process essentially involves performance measurement and evaluation. Performance measurement and evaluation involves determining the extent by which the actualized results overlap with the predefined goals of the projects.

The actions required for the realization of goals must be implemented through mobilization of resources and following a disciplined and organized approach, and a close coordination must be ensured among all responsible persons to achieve success.

Performances are meaningful if they are measurable and monitorable. Therefore, performances must not be difficult and expensive to monitor in practice. Too many or hard-to-understand performance measures must not be selected. Performance measures are instruments that ensure the measurement of the success of implementation results in particular. A performance is expressed in terms of potential, time, quality, and cost to ensure its measurability.

Deviations from goals must be identified and the reasons for these deviations must be analyzed. Corrective actions must be decided. Based on the results obtained, necessary revisions must be made to performance assessments, budget, actions, and goals.

Once implementations are up and running for a time, actual results that are achieved should be audited. In short, the actions described in the preceding paragraphs can be executed as follows and Table 8.2 can be used for auditing the results. First, gather the actual cost data. These will typically be available from records relating to the installation. The actual savings data should be available from company records, and should correspond to the analysis done during the project. Intangible benefits noted should result from a survey of the users, the people who are affected by the changes involved.

Next, record the actual results. Rate each using the same vowel letters used in step 5. In the case example, the results in terms of savings far exceeded expectations.

Table 8.2 Auditing Project Results—Example Case

Prepared by: _____ . Business/Problem: _Invoice system_ .

Authorized by: _____ . Project: _Accounting system change_ Date: _____ .

NO.	DESCRIPTION	MEASUREMENT	CURRENT	TARGET
1	Processing Errors Reduced	Errors/Month	2	0
2	Accts. Payable Overtime Reduced	Ave. Hours/Week	2	0
3	Morale Improved	Transfer Requests	2	0
4	System Streamlined	Minutes/Invoice	4	1
5	Supplier Relations Improved	Complaints/Month	1	0
6	Cycle Time to Process Reduced	Days to Process	2	1
7	Payments Simplified	Checks/Week	5	1

RESULTS AUDIT: _April/23_ | **RESPONSIBLE:** _RS_

INV. COSTS: PLANNED _$169,000.00_ **ACTUAL** _$167,179.20_ | **ADDITIONAL NOTES**

NET POT. SAV. PLANNED _$250,000_ **ACTUAL** _$371,392.02_

OTHER BENEFITS _Housekeeping & morale_

NO.	ACTUAL	RATING	COMMENTS
1	0	A	
2	0	A	
3	0	A	
4	0	A	
5	0	A	
6	25	E	
7	0	A	Elec. fund transfer
8			

In addition, Table 8.3 can be used to summarize the status of the projects with respect to the targeted goals.

While monitoring the performances and effects of the projects, we should always ask whether we have reached the set program target or not. If the answer is not sufficient to serve the main target, then analysis, ideas, and/or implementation plans should be considered again and accordingly additional improvements should be made. Figure 8.3 shows how to monitor the status of the total improvement target.

Table 8.3 Project Log—Example Case

PROJECT	RESPONSIBLE PERSON	START	COMPLETE	% TIME STATUS	% INVESTMENT COST STATUS	% POTENTIAL SAVINGS STATUS
Fit-up Problem—Redesign Bottom Plate of Model 1206	HB	4/20	6/20	100	110	90
Capacity Problem—Setup Time—#2 Shear	YO	7/13	8/13	100	90	85
Invoice System—Changes to Accounting System	RS	7/17	7/30	85	96	145

Status of gap closing per unit

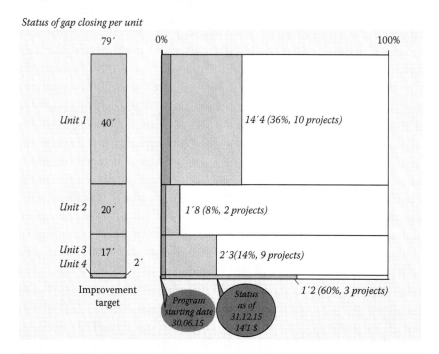

Figure 8.3 Status of total improvement target—example case.

Appendix

A.1 Sample Application

The following sample application follows the pattern of a simplified version of Systematic Improvement Planning (SIP) and some of its techniques. Though the sample application does not cover all the steps with its related techniques of a typical SIP, it still reflects the basics. The intention here is to support a better understanding of SIP by implementing its steps on a sample case in addition to the various examples introduced in the previous chapters.

The selected business is a mid-sized logistics firm that operates several warehouses and its main goal is to increase the operational performance and profitability of one of its strategic warehouse and exporting the successful results as reference points to its other facilities.

In the following sections, we assume the first two steps of SIP, which are Orientation and Investigating Current Status by Data Analysis and Clarifying Improvement Targets, have already been applied on the sample application and accordingly the following steps are put into action.

A.1.1 Determining Operational Problems
     by the Work Sampling Analysis Technique

Several measurement and analysis techniques (which are explained in Chapter 5) can be used individually or concurrently for determining operational problems and therefore leverages. However, for the sake of showing how any one of these techniques can be implemented on a real-life case, the work sampling technique is selected as an example.

A.1.1.1 Setting a Work Sampling Program and Orientation
To use the work sampling technique, we need to take into consideration the following issues:

- Work sampling program, as shown in Table A.1
- Determination of the activity areas and people that measurements will be done in and on, as shown in Table A.2
- Determination of the number of measurements and their frequencies, as shown in Table A.3
- Activities, activity areas, and duties lists, as shown in Tables A.4 to A.6
- Preparation of the form and measurement devices, as shown in Figure 5.11
- Organization of the measurement teams and leaders and their trainings
- Piloting and final modifications

A.1.1.2 Reports and Control of Documentation
After the observations are recorded on the Work Sampling Form (see Figure 5.11) according to the program that is set in preceding section, we are at the point of generating reports either by Excel® or a custom-made database software. An example report may look like the diagram shown in Figure A.1.

Table A.1   Work Sampling Program—Example

Prepared by	:		Rev. No		:	
Date	:		Page No		:	

NO	RESPONSIBLE	ACTION	30.06 – 13.07	14.07 – 27.07	28.07 – 10.08	11.08 – 24.08	25.08 – 07.09	08.09 – 21.09	22.09 – 05.10	06.10 – 19.10	20.10 – 02.11	03.11 – 16.11	17.11 – 30.11	01.12 – 14.12	15.12 – 28.12	29.12 – 11.01	12.01 – 25.01
			(shaded) = Planned											(solid) = Actual			
1		Determination of the activity areas and people that measurements will be done in and on	P														
2		Determination of the number of measurements and their frequencies	P														
3		Determination of activities, activity areas and duties lists	P	P													
4		Preparation of the form and measurement devices	P														
5		Organization of the measurement teams and leaders and their trainings			P												
6		Piloting and final modifications				P											
7		Measurements							P	P	P						
8		Recording measured values into tables							P	P	P						
9		Reporting and documentation control												P			
10		Analysis of measurements and determination of problems													P		
11		Grouping problems and prioritization															P
12																	

NOTES:	

Table A.2 Determination of Activity Areas and People Measured—Example

ACTIVITY AREAS	ROLES	TOTAL EMPLOYEES	OBSERVED EMPLOYEES	OBSERVERS
Picking	Picking operator	2	2	4
	Order filling labor	12	6	4
Receiving Area and	Receiving personnel	1	1	4
Defective Materials Area	Receiving dock personnel	2	2	4
	Receiving labor	1	1	4
	Receiving operator	8	3	4
Order Filling and Dock	Control personnel	1	1	4
	Loading labor	1	1	4
	Loading personnel	2	1	4
	Loading operator	4	2	4
	Order filling personnel	6	3	4
	Order filling operator	10	3	4

Table A.3 Determination of the Number of Measurements and Their Frequencies—Example

Population	6.5 hrs.*3 wks.*2 months*60	
Confidence interval	6%	Measurements in 1st, 2nd, 4th weeks
Confidence level	90%–95%–99%	Measurements on Tuesday, Wednesday, Thursday
Sample size	175–240–380	Measurements b/w July–September
Applied sample size	190	Measurements b/w 08.00–16.00 hours
Frequency	Every 15 min	
	28 observations/day	

A.1.1.3 Analysis of Measurements and Determination of Problems

Here we analyze the generated reports by conducting workshops and as a conclusion we may end up with several problem definitions by each group such as the ones that are shown in Table A.7.

A.1.1.4 Grouping Problems and Prioritization

We may assume that brainstorming sessions and using the tools that are explained in Chapter 5 will reveal categorized and prioritized leveraged problems such as the ones shown in Table A.8.

Table A.4   Activities List—Example

A	Equipment control and cleaning
B	Moving to storage zone
C	Picking materials
D	Transporting materials
E	Placing materials
F	Documentation work
G	Replacing batteries
H	Packaging
I	Receiving work orders
J	Active control
K	Quality control
L	Labeling
M	Counting materials
N	Parking forklifts and preparing them to use
O	Additional work
P	Training
Q	Being in meeting
R	Not working
S	Personal needs
T	Waiting assignments/work

Table A.5   Activity Areas List—Example

1	Supporting Materials Area
2	Picking Area
3	Office
4	Receiving Area
5	Blocking Area
6	Rejected Materials Area
7	Dock
8	Order Filling Area
9	Defective Materials Area
10	Recycling
11	Forklift Parking Area

A.1.2 Developing Alternative Improvement Ideas

A.1.2.1 Generation of Ideas

Through the use of alternative idea generation approaches that are explained in Chapter 6, for each determined and selected operational leveraged problem, we may generate several improvement ideas. Here, for the sake of simplicity, we demonstrate alternative improvement

Table A.6 Duties List—Example

PICKING OPERATOR

1	Fulfilling the requirements of the relevant activities documented by Order Filling teams
2	Transporting materials from Receiving Area to Picking Area
3	Transporting recycled materials to Picking Area
4	Controlling the temperatures and humidity rates of warehouse zones and recording them
5	Controlling Picking Area with respect to first in-first out (FIFO) criteria
6	Supporting loading, picking, and filling teams
7	Organizing Picking Area
8	Transporting empty pallets from Picking and Order Filling Areas to Pallet Storage Area
9	Transporting materials to Production Area
10	Maintaining and informing daily warehouse problems

ORDER FILLING LABOR

1	Collecting materials from the Picking Area
2	Packing prepared materials and transferring them
3	Stamping customer labels
4	Organizing and cleaning Picking Area
5	Informing equipment breakdowns to repair shop
6	Cleaning equipment
7	Picking materials

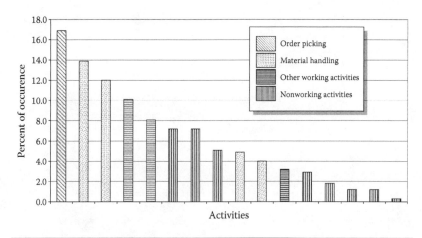

Figure A.1 Visualization of work sampling results—example.

Table A.7 Determined Problems—Example Group Work

1	MANAGERIAL ISSUES
	Implementation differences of the same role descriptions create difficulties for managers in organizing, controlling, and evaluating their job assignments
2	TRAINING
	Systematic and periodical training needs of management about managerial issues and warehouse employees about equipment and operations
3	LAYOUT PLANS
	Losses in space utilization, time, and expenditures due to the inappropriate use of methodologies and techniques
4	EQUIPMENT AND HARDWARE
	Inefficiencies in operations and loss in time due to the lack of and inappropriate setup of warehouse equipment and IT infrastructure

Table A.8 Prioritized Problems—Example

1	LAYOUT PLANS
	Losses in space utilization, time, and expenditures due to the inappropriate use of methodologies and techniques
2	TECHNOLOGY
	Picking, counting, addressing, … problems due to the inappropriate use of technology
3	MANPOWER LOSS
	Manpower utilization and planning problems in warehouse operations

Table A.9 Generated Ideas—Example

PROBLEM: MANPOWER UTILIZATION AND PLANNING PROBLEMS IN WAREHOUSE OPERATIONS

IDEA	DESCRIPTION
1	Approval of the materials should be done according to similar procedures and defined criteria, and be directly located in the warehouse zones.
2	Load the vehicle from the closest dock by the assigned operator.
3	Reorganize the roles and procedures with respect to operations and their systems.
4	Organize the ordering system, share the information, and guide the operations with respect to shipment plans.

ideas that were generated for only one leveraged problem (problem #3) such as the ones that are shown in Table A.9.

A.1.2.2 Description of Ideas

Here, for the sake of simplicity, a description is provided for only one of the generated ideas and at the conceptual level (idea #4) such as the one shown in Figure A.2. In reality, descriptions should be provided for all the generated ideas of every leveraged problem, not only at the conceptual level but in quantitative terms as well (such as

Problem:	Manpower utilization and planning problems in warehouse operations		
Idea Title:	Organize the ordering system, share the information and guide the operations with respect to shipment plans	Number:	
Person Responsible:			Date:

Description of Idea:
Decreasing excessive inventory levels
Increasing the quantity of highly circulated materials and decreasing the materials handling among zones
Preparing the availability of materials to be shipped by organizing the warehouse operations accordingly
Increasing the performance of FIFO approach
Decreasing the waiting lines of trucks by uniforming their arrival rates

Positive	**Negative**
Lowering the quantity of workers	*Difficulty in developing high-quality workers*
Increasing the availability of materials in inventory	*Difficulty in managing the change*
Keeping production and shipping plans at the required level	
Decreasing the transportation cost	
Potential:	**Cost of Implementation:**
Low—Medium	*Low—Medium*

Figure A.2 Description of an idea—example.

the one shown in Figure 6.4), and finally all the described ideas of every leveraged problem should be listed (such as the one shown in Table 6.1).

A.1.3 Evaluating Improvement Ideas

Evaluation of the improvement ideas of each leveraged problem should be done separately. Here, for the sake of simplicity, evaluation is applied on the ideas for problem #3 only by using a simple technique such as the one shown in Figure A.3. The ones that have a high potential and low cost are selected. More than one selected improvement idea may be selected for each leveraged problem. Finally, all the selected ides for all leveraged problems should be listed and prioritized (such as the one shown in Table 7.3).

A.1.4 Project Implementation Plan

A typical Installation Schedule for any selected idea is shown in Table A.10. For the given sample application, the controlling part of this step is not included.

A.2 Brainstorming

Broadly defined, brainstorming (Figure A.4) is a type of meeting arranged to produce creative ideas (Fligor 1990). The ideas generated are intended to reach a solution and to support the decision-making process. Brainstorming is actually a participative and practical technique commonly used to let organizations develop ideas.

The intention of brainstorming is to create a radical thinking environment, to adopt simple and mutual systematic thinking, to examine agreements and assumptions, and to reach a solution by consensus.

A.2.1 Preparatory Work

To manage brainstorming, it will be beneficial

- To reach a required solution by brainstorming, in other words, to reveal the intention of brainstorming
- To determine the sources of the results required to be reached

Problem: *Manpower utilization and planning problems in warehouse operations*

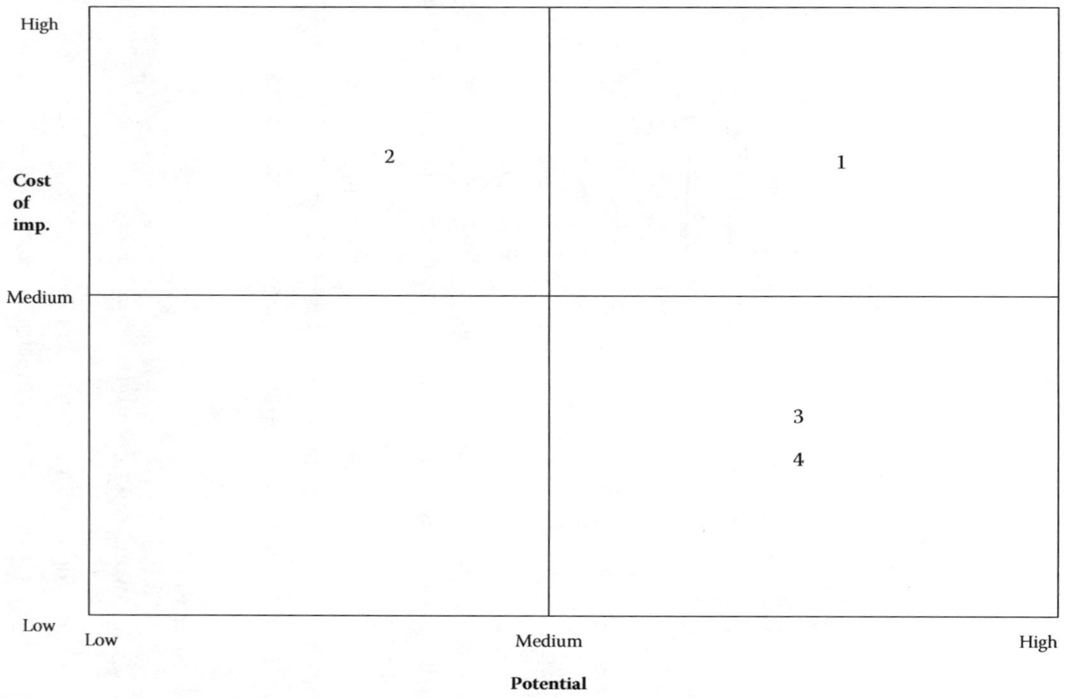

No	Title
1	*Approval of the materials should be done according to the similar procedures and defined criteria, and be directly located to the warehouse zones*
2	*Load the vehicle from the closest dock by the assigned operator*
3	*Reorganize the roles and procedures w.r.t. operations and their systems*
4	*Organize the ordering system, share the information and guide the operations w.r.t. shipment plans*

Figure A.3   Evaluating improvement ideas—example.

Table A.10 Installation Schedule—Example

Problem: *Manpower utilization and planning problems in warehouse operations*

Selected Idea: *Organize the ordering system, share the information and guide the operations w.r.t. shipment plans*

Prepared	:	
Date	:	
Rev. No	:	
Page No	:	

(■ = Planned; ■ (black) = Actual)

NO	RESPONSIBLE	ACTION	31.03 – 13.04	14.04 – 27.04	28.04 – 11.05	12.05 – 25.05	26.05 – 08.06	09.06 – 22.06	23.06 – 06.07	07.07 – 20.07	21.07 – 03.08	04.08 – 17.08	18.08 – 31.08	01.09 – 14.09	15.09 – 28.09		
1		*Compare determined maximum and minimum inventory levels with actual figures*	■														
2		*Determine the weekly palletized and SKU based sales figures*		■	■												
3		*During the shipment, use FIFO approach. Identify excess inventories and low demanded materials*				■											
4		*Eliminate the low demanded materials and ship them away*					■	■									
5		*Empty the excessive materials*							■	■							
6		*Schedule the shipment of the requested materials to the warehouses according to the production plans and reflect the probable changes to the shipment plans*									■	■	■	■	■		
7																	
8																	
9																	
10																	
NOTES:																	

- To understand well and define the problems to be solved
- To determine the developments and trends on the issues to be addressed

A.2.2 Pre-meeting

If possible, locations other than the office shall be selected for the meeting. Locations selected for brainstorming shall be indoors and isolated as much as possible. It will be beneficial to inform the individuals who will participate in the meeting beforehand regarding the issue and to give a specific period—such as one week—to have them think about it. Though this approach will enable the participants to generate more ideas during brainstorming, it can also cause the participants to affect each other by discussing the issue before brainstorming, so informing the participants beforehand shall be left to the decision of the meeting manager.

Figure A.4   Brainstorming meeting.

A.2.3 Brainstorming Meeting

Brainstorming meetings shall be held in a disciplined environment but not a formal one. The number of individuals attending the brainstorming meetings shall not be excessive and it shall be limited according to the purpose of brainstorming, meeting hall, and time to be allocated for the meeting. Participants with equal levels of knowledge and experience will be beneficial so everyone can contribute.

Brainstorming meetings shall be as active and lively as possible. It will be beneficial to have the brainstorming meetings managed by a meeting manager.

When starting the brainstorming meeting, even if the presence of attendants experienced in the subject area is not necessary it is recommended and will help the session succeed and reach the right solution. A recorder shall not be used during the brainstorming, as the participants would prefer it not to be. Everyone should actively participate in the sessions.

A.2.4 Start of the Meeting

Before starting brainstorming, the meeting manager shall clearly communicate to the participants the intention of the meeting, the result required, and the means by which the required result will be achieved.

The meeting manager shall enlighten the participants regarding the sections of the meeting and the length of each section. This will make the meeting more efficient. Adhering to the allotted time is the responsibility of the meeting manager.

For the positive start of brainstorming, the meeting manager should start with entertaining and interesting questions. At the beginning, a written expression of opinions is recommended to accelerate the adaptation of the participants.

A.2.5 Generating Ideas

Speaking opportunities should be given to the participants in a random manner. Here, the important thing is to allow everyone to express their opinions (the approach that is mostly recommended).

It is required that everyone shall speak in turn and present all of their opinions regarding the issue without being interrupted. A specific idea is selected and other ideas evoking the idea are encouraged.

Participants are required to express their ideas positively and then negatively (parallel thinking). As an extension of the parallel thinking approach, the six-hat method of Edward de Bono can be used. During the application of this method, it is expected that the participants wear various hats and make interpretations concerning the perspectives given by these hats. These perspectives are as follows: being objective, being emotional, being a risk taker, being positive, being a creative, and being judgmental.

During brainstorming, the meeting manager shall provide an environment for all the participants to speak. Here, the point not to be ignored is that participants may be silent because they are thinking, are shy, are mistrustful of the environment, or for other reasons. The meeting manager shall specify the ideas through repetition. Differences can arise in the second expression of an idea.

During the generation of ideas, the meeting manager shall not make any interpretations, criticisms, or evaluations. The meeting manager shall accept the ideas with maturity. A seemingly superficial idea can assist in the rise of an idea that will lead to a real result.

Ensuring that the issue is not diverted during the generation of ideas or getting a diverted issue back on track is the responsibility of the meeting manager. The meeting manager can ask questions such as *how, where, when,* and *why* to accelerate the meeting.

The meeting manager shall remind participants of the intention of the meeting during brainstorming. When the statement of the intention during the first brainstorming cannot assist in generation of ideas, a second statement of intent shall be selected.

A work session that becomes inefficient shall not be pushed to continue. It is beneficial to conclude the brainstorming sessions with the most unimagined ideas. This will assist the rise of different ideas.

During brainstorming, the meeting manager will number and write all ideas on a flip chart, using a nonerasable pen and in a way that attracts everyone's attention. All ideas shall be placed on the wall side to side so as to enable everyone to see them. During writing, it is important not to disturb the fluency of brainstorming and to use abbreviations to encourage the rise of new ideas.

A.2.6 Evaluation of Ideas

All the assembled ideas can be subject to elimination. To facilitate the process, the assembled list should be allocated to subgroups. It will be beneficial for the participants to perform the elimination.

During the elimination process, the meeting manager can use various techniques. For example, the ideas that are truly critical or that are impossible to evaluate or control can be kept and the others can be removed from the list according to the requirements of the meeting manager.

A.3 Working Forms

Copies of the forms used for Systematic Improvement Planning (SIP) are provided in this appendix. They may be used when solving your next improvement planning problem. You may reproduce the copies of these forms for your own use, provided you acknowledge their original source and keep their use within the copyright restrictions covering this book.

Each of the forms included in this section is explained in the text and is listed in its order of appearance in the text.

FORM CODE	FORM TITLE
RMA—756	Orientation Worksheet
IMECO—SIP—BS1	Business Status Worksheet
SIP—FI1	Financial-Improvement Target Diagram
SIP—TD1	Target Distribution Diagram
IMECO—SIP—P1	Classified Problems List w/Leverages
SIP—IL1	Identifying Improvement Leverages
SIP—ID1	Idea Description Worksheet
SIP—PA1	Idea Potential Aspects Worksheet
IMECO—SIP—AI1	Assessment of Ideas Summary Form
IMECO—SIP—QF1	Evaluation Form by Quantitative Factors
RMA—173	Evaluation of Alternatives Form
W. E. FILLMORE—PW1	Priority Worksheet
SIP—IS1	Implementation Schedule
IMECO—SIP—PR1	Project Results Worksheet
IMECO—SIP—PL1	Project Log
IMECO—SIP—IT1	Status of Total Improvement Target

Table A.11 Orientation Worksheet

Prepared by: _____ Business: _____

Authorized by: _____ Date: _____

PROGRAM TEAM   L-LEADER M-MEMBER R-RESOURCE O-OTHER | **HRS. REQD.**

NAME	ROLE	DEPARTMENT	PHONE	EST.	ACT.
			TOTAL		

PROGRAM ESSENTIALS

Objective(s)   _____
External Condition(s)   _____
Situation(s)   _____
Scope/Extent   _____

PLANNING ISSUES	Imp.	Resp.	Proposed Resolution	Ok by
1.				
2.				
3.				
4.				
5.				
6.				

Dominance/Importance Rating   Mark "X" if beyond control

PROGRAM SCHEDULE

NO.	ACTIVITY	WHO	HRS.REQD	DUE DATE	STATUS
1	Determination of Status				
2	Determination of Improvement Targets				
3	Get Facts, Analyze and Determine Leverages				
4	Challenge Alternative Solutions				
5	Option Evaluation				
6	Establish Implementation Plan and Control				

Reference Notes _____

Table A.12 Business Status Worksheet

Prepared by: _____ Business: _____

Authorized by: _____ Date: _____

	Factors	Status	Remarks for Target Setting
MARKETING/ SALES	Customer complaints		
	Customer due dates		
	Defect types and rates		
	Sales volumes		
	Products and their image		
	Effectivity of channels of distribution		
	After-sales service performance		
FIN./ ACC.	Financials and ratios		
	Economies of scale		
	Effectivity of cost control system		
OPERATIONS/ TECHNICAL	Process flows		
	Delays		
	Suppliers' performances		
	Suppliers' due dates		
	Raw materials cost and availability		
	Inventory turnover rates		
	Products and their qualities		
	Defect types and rates		
	Production volumes		
	Cycle times		
	Working hours, break times, variations in work load and peaks		
	Facility plans and materials handling		
	Facilities and machine capacity utilizations		
	Machine productivity		
	Machine maintenance efficiencies		
	Machine setup efficiencies		
	Appearance and cleanliness		
	Production planning and control system		
	Effectivity of operations control procedures		
	RandD, technology and innovation		
MANAGEMENT PERSONNEL	Strategies, values and policies		
	Organization structure		
	Management performance		
	Job descriptions and grades		
	Employee satisfaction		
	Employee turnover rates		
	Incentives to motivation		
	Employee trainings		
	Staffing levels		
	Labor capacity utilizations		
	Labor productivities		

+	Positive
	Neutral
-	Negative

FINANCIAL - IMPROVEMENT TARGET DIAGRAM

Prepared by:_____   Business: _____
Authorized by:_____   Date: _____

Firm vs. competitors

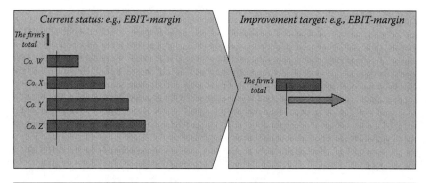

Figure A.5   SIP—FI1.

TARGET DISTRIBUTION DIAGRAM

Prepared by:_____   Business: _____
Authorized by:_____   Date: _____

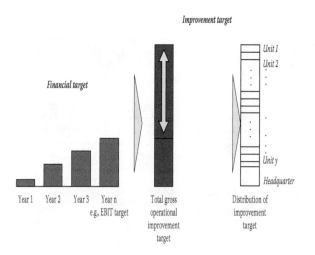

Figure A.6   SIP—TD1.

Table A.13 Classified Problems List with Their Leverages

Prepared by: _____ Business: _____

Authorized by: _____ Date: _____

NO	DESCRIPTION	TYPE (R – Random, C – Chronic)	LEVERAGES
1			• • • • •
2			• • • • •
3			• • • • •
4			• • • • •

NOTES

DECISION TREE APPROACH for IDENTIFYING IMPROVEMENT LEVERAGES

Prepared by:_____   Business: _____
Authorized by:_____   Date: _____

Decision tree breaks a problem down into smaller sub-problems to
identify improvement leverages

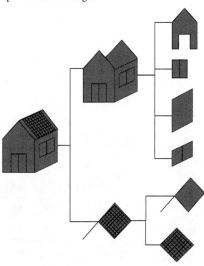

Figure A.7   SIP—IL1.

IDEA DESCRIPTION

Prepared by:_____   Business/Problem: _____
Authorized by:_____   Date: _____

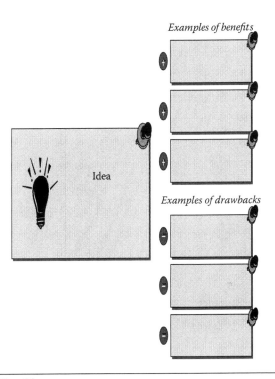

IDEA POTENTIAL ASPECTS

Prepared by:_____   Business/Problem: _____
Authorized by:_____   Date: _____

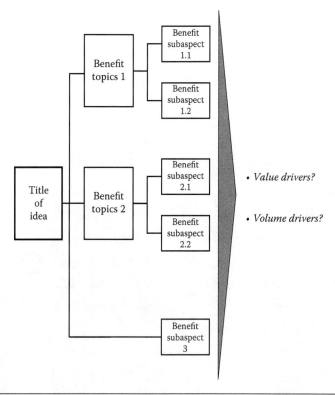

Table A.14 Assessment of Ideas Summary Form

Prepared by: _____ Business/Problem: _____

Authorized by: _____ Date: _____

IDEA	DESCRIPTION	REQUIREMENTS	IMPLEMENTATION COST	POTENTIAL
A				
B				
C				
D				
E				
IMPLEMENTATION COST: High (H), Medium (M), Low (L)			**POTENTIAL:** High (H), Medium (M), Low (L)	
NOTES				

NO	POTENTIAL ASPECTS	MEASURE (Value and/or Volume Drivers)	CURRENT	TARGET	A	B	C	D	E
1									
2									
3									
4									
5									
6									
7									

Table A.15 Evaluation Form by Quantitative Factors

	Bus.	No.
	By	Date

Ratings by .... Approved by ....

Problem:

Description of Alternative Improvement Ideas:
A.
B.
C.
D.
E.

	QUANTITATIVE FACTOR		ALTERNATIVE				
			A	B	C	D	E
1.	Annualized Net Savings (Sales Inc./Cost Red.)	c					
2.	Personnel						
3.	Material						
4.	Capital						
5.	Others						
6.							
7.	Annualized Investment Cost	a					
8.		b					
TOT.	Annualized Net Savings over Investment Cost						

Reference Notes:

a. Total Capital Investment/Expected Life (yrs.)
b. One-time costs don't need to be considered
c. Potential savings over differential implementation costs (yrs.)

d.
e.
f.

Table A.16 Evaluation of Alternatives Form

			Bus. _____ No. _____
EQUIVALENT WEIGHT			By _____ Date _____

Weight set by .... Tally by ....

Ratings by .... Approved by ....

EVALUATING DESCRIPTION

A = Almost Perfect, O = Ordinary Result
E = Especially Good, U = Unimportant Result
I = Important Result, X = Not Acceptable

Problem:

Description of Alternative Improvement Ideas:
X. _____
Y. _____
Z. _____
V. _____
W. _____

FACTOR/CONSIDERATION	WT.	ALTERNATIVE				
		X	Y	Z	V	W
1.						
2.						
3.						
4.						
5.						
6.						
7.						
8.						
TOT. Weighted Rate Down Total						

Reference Notes:

a. _____ d. _____
b. _____ e. _____
c. _____ f. _____

Table A.17 Priority Worksheet

Prepared by: _____ Business: _____

Authorized by: _____ Date: _____

NO.	DATE	IDEA - PROBLEM DESCRIPTION	IMPORTANCE	URGENCY	PRIORITY
1					
2					
3					
4					
5					
6					
7					
8					
9					
10					
11					
12					
13					
14					
15					

IMPORTANCE: Use A = Absolutely Necessary, E = Especially Important, I = Important, O = Ordinary Importance
URGENCY: Use a 1 to 10 scale; for example, 10 = Today, 7 = This week, 4 = This month, 1 = This year
PRIORITY: Convert Importance vowels to numbers (A=4, E=3, etc.) and multiply by Urgency values.

IMPLEMENTATION SCHEDULE

Prepared by:_____   Business/Problem: _____
Authorized by:_____   Project: _____ Date: _____

Installation schedule																									Notes and further action
No.	Action required	Who																							
1																									
2																									
3																									
4																									
5																									
6																									
7																									
8																									
9																									
10																									
11																									
12																									
13																									
14																									
15																									

Notes: _____

Figure A.10   SIP—IS1.

Table A.18 Project Results Worksheet

Prepared by: _____   Business/Problem: _____

Authorized by: _____   Project: _____   Date: _____

NO.	DESCRIPTION	MEASUREMENT	CURRENT	TARGET

RESULTS AUDIT:				RESPONSIBLE:
INV. COSTS: PLANNED ACTUAL				ADDITIONAL NOTES
NET POT. SAV.: PLANNED ACTUAL				
OTHER BENEFITS				
NO	ACTUAL	RATING	COMMENTS	

A = Almost Perfect,   O = Ordinary Result
E = Especially Good, U = Unimportant Result
I = Important Result, X = Not Acceptable

Table A.19 Project Log

Prepared by: _____ Business: _____

Authorized by: _____ Date: _____

PROJECT	RESPONSIBLE	START	COMPLETE	TIME % STATUS	INVESTMENT COST % STATUS	POTENTIAL SAVINGS % STATUS

STATUS of TOTAL IMPROVEMENT TARGET

Prepared by:_____   Business: _____

Authorized by:_____   Date: _____

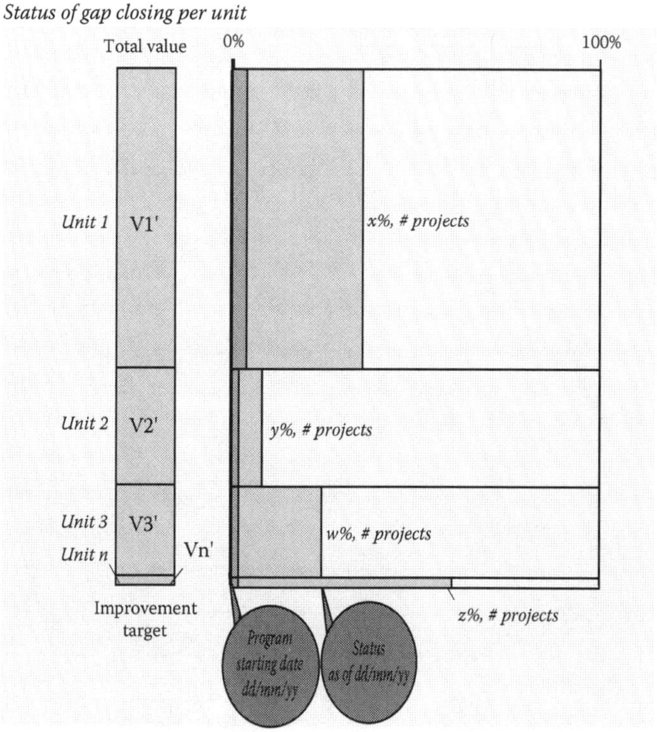

Status of gap closing per unit

Figure A.11   IMECO — SIP—IT1.

References

Barnes, R. M. 1980. *Motion and time study: Design and measurement of work.* New York: John Wiley & Sons.

Coelli, T. J., Rao, D. S. P., O'Donnell, C. J., and Battese, G. E. 2005. *An introduction to efficiency and productivity analysis.* New York: Springer Science+Business Media.

Fligor, M. 1990. *Brainstorming: The book of topics.* Storrs, CT: Creative Learning Press.

George, M. L., and Maxey, J. 2004. *The Lean Six Sigma pocket toolbook.* Columbus, OH: McGraw-Hill.

Hales, H. L., and Andersen, B. 2002. *Systematic planning of manufacturing cells.* Dearborn, MI: Society of Manufacturing Engineers.

Ham, I. 2013. *Group technology: Applications to production management.* New York: Springer Science+Business Media.

Harrington, H. J., and Esseling, K. 1997. *Business process improvement workbook.* Columbus, OH: McGraw-Hill.

Hayes, B. E. 2008. *Measuring customer satisfaction and loyalty.* Milwaukee, WI: ASQ Quality Press.

Keown, A. J., and Martin, J. 2001. *Financial management: Principles and applications.* Upper Saddle River, NJ: Prentice Hall.

Kerzner, H. R. 2013. *Project management: A systems approach to planning, scheduling, and controlling.* Hoboken, NJ: John Wiley & Sons.

Liker, J. 2004. *The Toyota way: 14 management principles from the world's greatest manufacturer.* Columbus, OH: McGraw-Hill.

Mogensen, A. 1932. *Common sense applied to motion and time study.* Columbus, OH: McGraw-Hill.

Monden, Y. 2011. *Toyota production system: An integrated approach to just-in-time.* New York: Productivity Press.

Muther, R. 1974. *Systematic layout planning.* Boston, MA: Cahners Books.

Muther, R. 2011. *Planning by design.* Kansas City, MO: Institute for High Performance Planners.

Muther, R., and Haganas, K. 1969. *Systematic handling analysis.* Kansas City, MO: Management and Industrial Research Publications.

Parmenter, D. 2010. *Key performance indicators: Developing, implementing, and using winning KPIs.* Hoboken, NJ: John Wiley & Sons.

Plenert, G. 2011. *Strategic continuous process improvement.* Columbus, OH: McGraw-Hill.

Scholl, A. 1999. *Balancing and sequencing of assembly lines.* Heidelberg: Physica-Verlag.

Shingo, S. 1986. *Zero-quality control: Source inspection and the Poka-Yoke system.* New York: Productivity Press.

Shingo, S. 1996. *Quick changeover for operators: The SMED system.* New York: Productivity Press.

Skinner, D. C. 2009. *Introduction to decision analysis.* Sugar Land, TX: Probabilistic Publishing.

Venge, B. 2011. *Continuous improvement manifesto—The ultimate guide to improving any business under any circumstance.* New York: Rockstone Media.

Index

Page numbers with f and t refer to figures and tables, respectively.